Introduction to Special Functions for Applied Mathematics

Introduction to Special Functions for Applied Mathematics introduces readers to the topic of special functions, with a particular focus on applications. Designed to build swiftly from the more basic special functions towards more advanced material, the book is ideally suited for an intensive one semester course. Complemented with various solved examples and exercises to support students and instructors, the book can be used for both self-study and directed learning.

Features

- Suitable for graduate level students or beginning PhD students in mathematics, physics, statistics, and economics
- No previous background in complex analysis required
- Numerous solved examples and exercises.

Daniele Ritelli is an Associate Professor of mathematical analysis at the Department of Statistical Sciences of the University of Bologna in Italy. He teaches ordinary and partial differential equations and special functions in various undergraduate and master's degree courses in statistics and quantitative finance. His research interests concern special functions, generalised trigonometric functions, ordinary differential equations, and applied mathematics.

Introduction to Special Functions for Applied Mathematics

Daniele Ritelli

CRC Press
Taylor & Francis Group
Boca Raton London New York

CRC Press is an imprint of the
Taylor & Francis Group, an **informa** business
A CHAPMAN & HALL BOOK

Designed cover image: Daniele Ritelli

First edition published 2026
by CRC Press
2385 NW Executive Center Drive, Suite 320, Boca Raton FL 33431

and by CRC Press
4 Park Square, Milton Park, Abingdon, Oxon, OX14 4RN

CRC Press is an imprint of Taylor & Francis Group, LLC

ISBN: 978-0-367-77469-1 (hbk)
ISBN: 978-0-367-77470-7 (pbk)
ISBN: 978-1-003-17154-6 (ebk)

DOI: 10.1201/9781003171546

Typeset in CMR10 font
by KnowledgeWorks Global Ltd.

Publisher's note: This book has been prepared from camera-ready copy provided by the authors.

Contents

Preface

With this book, we want to introduce students with a quantitative education, not necessarily advanced, to the study of Special Functions. This need arises from lecturing to PhD students in Statistics at Bologna University; their mathematical curriculum, following the Italian University syllabus, deals with differential and integral calculus in one or more variables without delving into more advanced subjects. On the other hand, when dealing with Statistics and Probability, especially in their computational aspects, one inevitably encounters numerous applications that require the operational use of various special functions to cite just some reference among many in the field, such as [59, 60, 67, 90]. Working with any mathematical or probabilistic model, one needs to deal with Eulerian functions, which we will present in chapters 2 and 3, Bessell functions, which we will discuss in chapters 7 and 8, hypergeometric functions, chapters 9 and 10. These tendencies not to cover advanced mathematical topics depends, in my opinion, on the fact that in the Italian system, University education in the field of Statistics is contiguous to that in the Social-Economic field, where the strategy, as I have been able to verify personally in over thirty years of lecturing, is that of limiting to provide basic mathematical instruments for the majority of the students, leaving to the (few) motivated scholars, with a view to a career in the academic world, the task of individually equipping themselves with advanced mathematics when needed. For these reasons, this textbook also seeks to be a textbook that can be used as a self-study tool. Of course, the applicative interest of the special functions does not exclude that this text can also be used as a primer by students of the physical, mathematical, and engineering sciences since, although many excellent treatises and textbooks are available in this field, they are often written assuming advanced prior knowledge, especially in complex analysis. Conversely, these lectures have been prepared to quickly bring those who need to possess these skills to a reasonable understanding of special functions. Naturally, I hope that this book intended as an initial introduction to the subject, will facilitate those interested in studying more advanced treatises and hopefully enlarge their mathematical horizon, but at the same time, enable those with interest in applied sciences to be able to work actively on models that employ the main special functions in their realisation.

Having explained this book's purpose, clarifying the meaning of special functions is equally important. Leaving aside the set definition of a function as a specific subset of the Cartesian product of two functions, see [62, page

30]; our attention will mainly be devoted to functions of one real or complex variable. But in what sense are the functions we will study in this book "special"? The adjective special is commonly used to indicate that the functions under study are not defined through algebraic operations, such as polynomial and rational and exponential functions or inversions. Trigonometric functions are the first example of special functions, although they are not commonly thought of as such since their geometric definition allows them to be defined at least intuitively. If, for example, we think about the geometric meaning of sine, cosine, and tangent, we understand that, on the one hand, the determination of the values of these functions is immediate for particular angular values such as 45 degrees (or $\pi/4$ radians). Still, at the same time, it is not that immediate to compute their values for closer values, such as 40 degrees (or $2/9\,\pi$ radians). This problem is solved by expressing trigonometric functions by their power series, which is made possible by the fact that they are solutions of specific ordinary differential equations with assigned initial conditions; e.g. the sine function is the solution of the initial value problem:

$$y'' + y = 0,\ y(0) = 0,\ y'(0) = 1\,. \tag{S}$$

In chapter 5, we will explain the technique that allows us to conclude that the following power series gives the solution to (S):

$$y = \sin x = \sum_{n=0}^{\infty} (-1)^n \frac{x^{2n+1}}{(2n+1)!}\,.$$

The functions we will introduce in chapters 6, 7, 8, 9, and 10 are also obtained by solving linear differential equations, which have variable coefficients. As one can imagine, the differential equations that identify these functions often come from applications of great importance, making knowledge of the functions that solve them essential. It frequently happens that a class of special functions originating from the solution of a problem finds applications in different contexts; for instance, the differential equation (7.1) on page 164, known as the Bessel equation, was initially used by Euler in models regarding the vibrations of a stretched membrane, then was employed by the astronomer F. Bessel to establish of the orbit of Halley's comet and since then Bessel functions have been used in countless applications. Ultimately, the functions presented in chapters 6, 7, 8, 9, and 10 are "special", as the power series identified by the differential equations that define them differ from the usual functions considered in Calculus courses: exponentials, logarithms, trigonometric functions, and geometric series. Instead, the Gamma and Beta functions of chapter 2 are defined by integrals depending on one or two parameters, where these integrals are not, except for particular values of the parameters being considered, expressible using the usual functions (polynomial, fractional, irrational, logarithmic, exponential, trigonometric, and their inverses). The Digamma function exposed in chapter 3 is defined as the logarithmic derivative of the Gamma function. In contrast, chapter 4 is dedicated to the Zeta function, which is introduced as a parametric infinite series. The final chapter collects exercises and applications that involve all the functions covered throughout the text to allow those who wish to deal with concrete applications to benefit from many examples. Furthermore, the proposed exercises are useful in fixing and assimilating the material covered. The notions necessary for understanding each element covered will be indicated. In this way, the reader can skip to the exercises that [s]he can carry out without completing the reading of the whole manuscript.

We assume the reader knows differential and integral calculus in one or more dimensions. The first chapter has a preparatory function by illustrating the results, usually outside the introductory calculus courses used during this presentation: Taylor and Fourier series, a primer in Complex Analysis, and Fourier and Laplace transforms. If the reader has already had the opportunity to deal with this topic, the chapter can easily be skipped. It is also possible to start reading from the second chapter and return to the contents of the first when needed. In this sense, the references in the text make it completely easy. This is how I behave when I present these contents in a lecture, as the first chapter works as the course toolbox. Chapter 5 also provides the tools for treating linear differential equations that generate the Special Functions explained in the following chapters. Anyone with knowledge on the subject can skip it. The exposition is traditional, using the pattern definition, theorem, proposition, lemma, and remark; notice that these items can be numbered; the first number is for the chapter, and the second is progressive; when a theorem, definition, or remark, it is not numbered it means that it will be not recalled later in the text.

In presenting the matter, I occasionally felt it necessary to make some brief historical mention, with the aim above all of illustrating the origin of the problems that led to the study of special functions starting from the conversations between Leonhard Euler, Daniel Bernoulli, and Christian Goldbach, between 1729 and 1730, which led to today's definition of the Gamma function. Naturally, also to keep the exposition within acceptable limits, these hints have no claim to be exhaustive but are offered with the hope of arousing the reader's curiosity, who can find a very in-depth treatment of these topics in the remarkable Jacques Dutka contributions [43, 41, 42, 40].

Since this is an introductory text, I have chosen to exclude some topics, both for reasons of exposition space and for the prerequisites assumed in the present exposition: the theory of orthogonal polynomials and Fourier Bessel series is only touched upon, while the multivariate hypergeometric functions, elliptic functions, and integrals have not found space in this introduction. However, this will allow those interested to access the specialist literature quickly.

Special thanks go to Antonio Edoardo Bargellini, who checked the manuscript several times. The responsibility for the imprecisions and typos that remain in the manuscript is entirely mine.

Chapter 1

Preliminaries

In this initial chapter, we quickly recall the mathematical concepts that will form the basis of our future discussion of special functions. In particular, we will discuss

- Sequences of functions
- Power series
- Complex analysis
- Fourier series
- Laplace and Fourier transforms

We present a brief review, and the themes we address here deserve to be explored in depth by scholars who, for whatever reason, must use advanced mathematical tools in their activity. Readers familiar with these subjects can easily skip and move to chapter 2. These topics are dealt with in countless manuals; here we point out, among many others, for power and Fourier series [112] chapters 4 and 14, [39] chapters 3 and 8, while for integral transforms, we indicate for Fourier transform [89, 72] and for Laplace transform [44, 107, 38]. Finally, although these lectures are prepared in an attempt to minimise the use of complex analysis, as they are addressed to an audience with no prior knowledge of the subject, we feel it is appropriate to decline the principal aspects of this theory, referring to the vast literature on the subject; here we merely cite, among many others, [7, 65, 92, 93], for the demonstrations of the statements we will present. Those interested in revisiting the notions presented in the following chapters with complex analysis tools will find the recent monographs by C. Viola [111] and B. Patra [92] very useful and informative.

1.1 Sequences of functions

Definition. A sequence of functions defined on the interval I is a numerable family $(f_n)_{n\in\mathbb{N}}$ of real (or complex) valued functions $f_n : I \to \mathbb{R}$ (or $\mathbb{C}$).

We say that (f_n) converges to f if, for any $x \in I$, there exists the limit of the sequence $(f_n(x))_{n\in\mathbb{N}}$ and its value is $f(x)$.

DOI: 10.1201/9781003171546-1

Convergence, which will be symbolised (as customary) by $f_n \to f$, can be reformulated in ε notation in the usual way: if for any $\varepsilon > 0$ and for any $x \in I$, there exists $n(\varepsilon, x) \in \mathbb{N}$ such that, for any $n \in \mathbb{N}$ with $n > n(\varepsilon, x)$:

$$|f_n(x) - f(x)| < \varepsilon.$$

The notion of convergence does not guarantee the possibility of performing the so-called exchange between limit and integration. We explain this fact by giving an example from [84].

Example. Let $f_n : [0, 1] \to \mathbb{R}$ defined as $f_n(x) = \dfrac{2^n\, x}{1 + n\, 2^n\, x^2}$. We see that $f_n \to 0$ since for $n \to \infty$ f_n behaves as $\dfrac{1}{n\, x}$. Computing the integral, we find

$$\int_0^1 f_n(x)\mathrm{d}x = \int_0^1 \frac{2^n\, x}{1 + n\, 2^n\, x^2}\mathrm{d}x = \frac{1}{2n} \ln\left(1 + n\, 2^n\right).$$

Hence

$$\lim_{n\to\infty} \int_0^1 f_n(x)\, \mathrm{d}x = \lim_{n\to\infty} \frac{1}{2n} \ln\left(1 + n\, 2^n\right) = \frac{1}{2} \ln 2,$$

while

$$\int_0^1 \lim_{n\to\infty} f_n(x)\, \mathrm{d}x = \int_0^1 \lim_{n\to\infty} 0\, \mathrm{d}x = 0\,.$$

Thus, in this case:

$$\lim_{n\to\infty} \int_0^1 f_n(x)\, \mathrm{d}x \neq \int_0^1 \lim_{n\to\infty} f_n(x)\, \mathrm{d}x.$$

This example suggests the importance of establishing appropriate conditions that allow the limit and integral operations to be interchanged, unlike what we have just seen. Passing to the limit can also cause continuity to be lost.

Example. Let $f_n : [0,\, 1] \to \mathbb{R}$ defined as $f(x) = x^n$. Each f_n is continuous, but

$$f(x) = \lim_{n\to\infty} f_n(x) = \begin{cases} 0 & \text{if } 0 \le x < 1, \\ 1 & \text{if } x = 1, \end{cases}$$

it is not continuous.

Definition. If (f_n) is a sequence of functions defined on I, we say that f_n converges **uniformly** to f if, for any $\varepsilon > 0$, there exists $n_\varepsilon \in \mathbb{N}$ such that, for $n \in \mathbb{N}$, $n > n_\varepsilon$:

$$\sup_{x\in I} |f_n(x) - f(x)| < \varepsilon.$$

Uniform convergence implies convergence. The converse does not hold. Uniform convergence is ensured if the following Cauchy condition is met; for the proof, see [101], Theorem 2.9, page 16.

Theorem. The following statements are equivalent:

(i) (f_n) converges uniformly;

(ii) for any $\varepsilon > 0$, there exists $n_\varepsilon \in \mathbb{N}$ such that if $n, m > n_\varepsilon$, $n, m \in \mathbb{N}$, then:

$$|f_n(x) - f_m(x)| < \varepsilon, \qquad \text{for any } x \in I.$$

Uniform convergence has remarkable properties. The first concerns continuity.

Theorem. If (f_n) is a sequence of continuous functions on a closed and bounded interval $[a, b]$, which converges uniformly to f, then f is a continuous function.

Proof. We refer, for instance, to [101] Theorem 2.14, pages 19–20. □

Uniform convergence allows the exchange between limit and integral when dealing with sequences of continuous functions defined on compact intervals.

Theorem 1.1. If (f_n) is a sequence of continuous functions on $[a, b]$, converging uniformly to $f(x)$, then

$$\lim_{n\to\infty} \int_a^b f_n(x)\mathrm{d}x = \int_a^b \lim_{n\to\infty} f_n(x)\mathrm{d}x\,.$$

Proof. For the proof, we refer to [101] Theorem 2.15, page 20. □

Concerning the derivation operator, the uniform convergence of a sequence implies the uniform convergence of the sequence of its derivatives.

Theorem 1.2. Let (f_n) be a sequence of continuous functions on $[a, b]$. Assume that each f_n is differentiable, with continuous derivative, and that:

(i) $\lim_{n\to\infty} f_n(x) = f(x)$, for any $x \in [a, b]$

(ii) (f_n') converges uniformly in $[a, b]$.

Then $f(x)$ is differentiable and

$$\lim_{n\to\infty} f_n'(x) = f'(x).$$

We also recall a result, due to U. Dini, and the remarkable Corollary 1.4, very useful in many applications; for their proof, we refer to [39, page 38].

Theorem 1.3. Let (f_n) be a sequence of continuous functions, converging to a continuous function f, on the interval $[a, b]$. Furthermore, assume that, for any $x \in [a, b]$ and for any $n \in \mathbb{N}$, $f_n(x) \geq f_{n+1}(x)$. Then f_n converges uniformly to f in $[a, b]$.

Remark. In Theorem 1.3, hypothesis $f_n(x) \geq f_{n+1}(x)$ can be replaced reverting the monotonicity assumption, i.e. assuming $f_n(x) \leq f_{n+1}(x)$.

Corollary 1.4. Let (f_n) be a sequence of nonnegative, continuous, and integrable functions, defined on $\mathbb{R}$, and assume that it converges pointwise to f, which is also nonnegative, continuous, and integrable. Suppose further that it is either $0 \leq f_n(x) \leq f_{n+1}(x) \leq f(x)$ or $0 \leq f(x) \leq f_{n+1}(x) \leq f_n(x)$, for any $x \in \mathbb{R}$ and any $n \in \mathbb{N}$. Then:

$$\lim_{n\to\infty} \int_{-\infty}^{+\infty} f_n(x) \mathrm{d}x = \int_{-\infty}^{+\infty} f(x) \mathrm{d}x.$$

There may be situations where the hypothesis of uniform convergence is not satisfied, but the passage to the limit remains possible if appropriate assumptions are verified. The following theorem, due to Arzelà, [15], dating back to 1885, ensures the passage to the limit using the uniform boundedness of the sequence of functions; we address to the interesting didactic article [78] and to [63, chapter 5].

Theorem 1.5. Let (f_n) be a sequence of integrable functions defined on $[a, b]$, bounded and closed interval convergent to f on $[a, b]$, being f integrable. Moreover, assume that $|f_n(x)| \leq M$, being $M > 0$. Then

$$\lim_{n\to+\infty} \int_a^b |f_n(x) - f(x)| \, \mathrm{d}x = 0.$$

In particular, we have the following:

$$\lim_{n\to+\infty} \int_a^b f_n(x) \, \mathrm{d}x = \int_a^b \lim_{n\to+\infty} f_n(x) \, \mathrm{d}x = \int_a^b f(x) \, \mathrm{d}x.$$

1.2 Series of functions

The transformation of a sequence of real (complex) numbers into an infinite series is the same as transforming sequences into a series of functions.

Definition. We say that the series of functions

$$\sum_{n=1}^{\infty} f_n(x) = f_1(x) + f_2(x) + \cdots + f_m(x) + \cdots\cdots \tag{1.1}$$

converges if the sequence of its partial sums $s_n(x) = \sum_{k=1}^{n} f_k(x)$ converges. When the convergence is uniform, we say that series (1.1) converges uniformly.

Remark. Introducing the remainder $r_n(x) := f(x) - f_n(x)$, then (1.1) converges uniformly for $x \in [a, b]$ if, for any $\varepsilon > 0$, there exists n_ε such that:

$$\sup_{x\in[a,b]} |r_n(x)| < \varepsilon, \qquad \text{for any } n > n_\varepsilon.$$

When the series

$$\sum_{n=1}^{\infty} |f_n(x)|$$

converges, its convergence may not be uniform. A condition that allows us to deduce the uniform convergence of a series of functions by studying a suitable numerical series is due to K. Weierstrass, known as the Weierstrass M-test.

Theorem. Let (f_n) be a sequence of functions defined on $[a, b]$. Assume that for any $n \in \mathbb{N}$, there exists $M_n \in \mathbb{N}$ such that $|f_n(x)| \leq M_n$ for any $x \in [a, b]$. Moreover, assume that:

$$\sum_{n=1}^{\infty} M_n$$

converges, then (1.1) converges uniformly in $[a, b]$.

Proof. See [101] Theorem 2.27 therein. □

Extending the Theorems 1.1 and 1.2 to the case of a series of functions is straightforward.

Theorem 1.6. If series (1.1) converges uniformly on $[a, b]$, then:

$$\sum_{n=1}^{\infty} \int_a^b f_n(x) \mathrm{d}x = \int_a^b \sum_{n=1}^{\infty} f_n(x)\, \mathrm{d}x. \tag{1.2}$$

Theorem 1.7. Assume that (1.1) converges uniformly on $[a,\, b]$, and assume that $f'_n(x)$ exists and it is continuous for any $x \in [a,\, b]$. Assume further that the series of derivatives is uniformly convergent. If $f(x)$ denotes the sum of the series (1.1), then $f(x)$ is differentiable and, for any $x \in [a\,, b]$:

$$f'(x) = \sum_{n=1}^{\infty} f'_n(x)\,.$$

The derivatives at the boundary of the interval are understood as right and left derivatives.

In the case of sequences of positive functions, the integration by series is correct by monotone convergence; referring, for instance, to [63, page 175], we can state the following:

Theorem 1.8. If (f_n) is a sequence of positive continuous functions on $[a, b]$, then the exchange between the series summation and the integral, equation (1.2), holds.

In the case of sequences with alternate terms, we state a necessary condition that allows the use of Arzelà's Theorem 1.5, which leads to the following result.

Theorem 1.9. Assume that (f_n) is a sequence of positive, continuous, and decreasing functions on $[a, b]$. Assume that

$$\sum_{n=1}^{\infty}(-1)^{n-1} f_n(x)$$

is a continuous function on $[a, b]$, then

$$\sum_{n=1}^{\infty}(-1)^{n-1} \int_a^b f_n(x)\mathrm{d}x = \int_a^b \sum_{n=1}^{\infty}(-1)^{n-1} f_n(x)\,\mathrm{d}x.$$

It is worth noting that the thesis follows from the fact that n-th partial sum $s_n(x)$ is nonnegative; moreover, since $s_n(x) \leq f_1(x)$, we can invoke Arzelà's theorem.

We state a final useful result on the integration by series in the case of absolute convergence.

Theorem. Assume that (f_n) is a sequence of continuous function on $[a, b]$ and assume that $g : [a, b] \to \mathbb{R}$ is a positive continuous function such that, for any $x \in [a, b]$, the following inequality

$$\sum_{n=1}^{\infty} |f_n(x)| \leq g(x)$$

is satisfied. If, moreover:

$$\sum_{n=1}^{\infty} f_n(x)$$

is a continuous function on $[a, b]$, then (1.2) holds.

1.3 Power series

Definition. Given the sequence $(a_n)_{n \geq 0}$ and $x_0 \in \mathbb{R}$, the function:

$$f(x) = \sum_{n=0}^{\infty} a_n \,(x - x_0)^n \tag{1.3}$$

is said power series,

We need to describe the set of values $x \in \mathbb{R}$ so that (1.3) converges uniformly.

Remark. It is not restrictive, after a translation, to consider power series with $x_0 = 0$.

$$\sum_{n=0}^{\infty} a_n x^n. \tag{1.4}$$

We state the main properties of the power series

Theorem. If (1.4) converges for $x = r_0$ then, for any $0 \le r < |r_0|$, it is absolutely and uniformly convergent in $[-r, r]$.

The fundamental result on the power series is due to Cauchy and Hadamard.

Theorem (Cauchy–Hadamard). For the power series (1.4), only one of the following alternatives holds:

(i) series (1.4) converges for any x;

(ii) series (1.4) converges only for $x = 0$;

(iii) there exists a positive number r such that series (1.4) converges for any $x \in]-r, r[$ and diverges for any $x \in]-\infty, -r[\cup]r, +\infty[$.

The interval within which (1.4) converges is called the interval of convergence and r is called the convergence radius.

The radius of convergence can be calculated as follows.

Theorem. Consider the power series (1.4) and assume that the following limit exists:

$$\ell = \lim_{n\to\infty} \left| \frac{a_{n+1}}{a_n} \right| .$$

Then:

(i) if $\ell = \infty$, series (1.4) converges only for $x = 0$;

(ii) if $\ell = 0$, series (1.4) converges for all x;

(iii) if $\ell > 0$, series (1.4) converges for $|x| < \dfrac{1}{\ell}$.

The convergence radius of (1.4) is then $r = \dfrac{1}{\ell}$.

Example. The following power series (1.5), known as the geometric series, has convergence radius $r = 1$.

$$\sum_{n=0}^{\infty} x^n. \tag{1.5}$$

In fact, since in (1.5), it is $a_n = 1$ for all $n \in \mathbb{N}$, thus:

$$\lim_{n\to\infty}\left|\frac{a_{n+1}}{a_n}\right| = 1.$$

Hence series (1.5) converges for $-1 < x < 1$. At the boundary of the interval of convergence, i.e. $x = 1$ and $x = -1$, series (1.5) does not converge. In conclusion, the interval of convergence of (1.5) is the open interval $(-1, 1)$.

Remark. Power series (1.6) has a radius of convergence $r = 1$.

$$\sum_{n=1}^{\infty} \frac{x^n}{n}. \tag{1.6}$$

In fact:

$$\lim_{n\to\infty}\left|\frac{a_{n+1}}{a_n}\right| = \lim_{n\to\infty}\frac{n}{n+1} = 1 .$$

that is, (1.6) converges for $-1 < x < 1$.

At the boundary of the interval of convergence, (1.6) behaves as follows: when $x = 1$, it reduces to the divergent series:

$$\sum_{n=0}^{\infty} \frac{1}{n},$$

while, when $x = -1$, (1.6) reduces to the convergent series:

$$\sum_{n=0}^{\infty} (-1)^n \frac{1}{n}.$$

The interval of convergence of (1.6) is, thus, $[-1\,, 1)$.

Remark. Series (1.7), given below, has an infinite radius of convergence:

$$\sum_{n=0}^{\infty} \frac{x^n}{n!}. \tag{1.7}$$

In fact:

$$\lim_{n\to\infty}\left|\frac{a_{n+1}}{a_n}\right| = \lim_{n\to\infty}\frac{1}{n+1} = 0.$$

It is possible to differentiate and integrate power series, as stated in the following Theorem 1.10, which we include for completeness, as it represents a particular case of Theorems 1.6 and 1.7.

Theorem 1.10. Let $f(x)$ be the sum of the power series (1.4), with radius of convergence r. The following results hold.

(i) $f(x)$ is differentiable and, for any $|x| < r$, it is:

$$f'(x) = \sum_{n=1}^{\infty} n\, a_n\, x^{n-1}\,;$$

(ii) if $F(x)$ is the *primitive* of $f(x)$, which vanishes for $x = 0$, then:

$$F(x) = \sum_{n=0}^{\infty} \frac{a_n}{n+1} x^{n+1}\,.$$

The radius of convergence of $f'(x)$ and $F(x)$ is that of $f(x)$.

Theorem. Consider two power series with radii of convergence r_1 and r_2, respectively:

$$f_1(x) = \sum_{n=0}^{\infty} a_n x^n, \qquad f_2(x) = \sum_{n=0}^{\infty} b_n x^n\,. \tag{1.8}$$

Then $r = \min\{r_1\,, r_2\}$ is the radius of convergence of

$$(f_1 + f_2)(x) = \sum_{n=0}^{\infty} (a_n + b_n)\, x^n\,.$$

If $\alpha \in \mathbb{R}$, then r_1 is the radius of convergence of

$$\alpha\, f_1(x) = \sum_{n=0}^{\infty} \alpha\, a_n\, x^n.$$

We state, without proof, the following Theorem 1.11, due to Cauchy, concerning the product of two power series.

Theorem 1.11. Consider again the power series in (1.8) and define

$$c_n = \sum_{j=0}^{n} a_j\, b_{n-j}.$$

Then

$$\sum_{n=0}^{\infty} a_n\, x^n \sum_{n=0}^{\infty} b_n\, x^n = \sum_{n=0}^{\infty} c_n\, x^n. \tag{1.9}$$

Series (1.9) has the interval of convergence given by $|x| < r = \min\{r_1, r_2\}$, and its sum is the pointwise product $f_1(x)\, f_2(x)$.

Remark. Notice that

$$\begin{aligned}
c_0 &= a_0\, b_0\,, \\
c_1 &= a_0\, b_1 + a_1\, b_0\,, \\
&\vdots \\
c_n &= a_0\, b_n + a_1\, b_{n-1} + \cdots + a_{n-1}\, b_1 + a_n\, b_0.
\end{aligned}$$

1.4 Taylor–MacLaurin series

Our starting point, here, is the Taylor formula with Lagrange remainder.

Theorem 1.12. Let $f : I \to \mathbb{R}$ be a function that admits derivatives of any order at $x_0 \in I$. If $x \in I$, then there exists a real number ξ, between x and x_0, such that:

$$\begin{aligned} f(x) &= P_n\,(\,f(x)\,,\,x_0\,) + R_n\,(\,f(x)\,,\,x_0\,) \\ &= \sum_{k=0}^{n} \frac{f^{(k)}(x_0)}{k!}(x-x_0)^k + \frac{f^{(n+1)}(\xi)}{(n+1)!}(x-x_0)^{n+1}. \end{aligned} \tag{1.10}$$

If

$$\lim_{n\to\infty} R_n\,(f(x), x_0) = \lim_{n\to\infty} \frac{f^{(n+1)}(\xi)}{(n+1)!}(x-x_0)^{n+1} = 0,$$

then

$$f(x) = \sum_{n=0}^{\infty} \frac{f^{(n)}(x_0)}{n!}(x-x_0)^n. \tag{1.11}$$

An operating condition useful in applications is seen in [101] Corollary 2.48 for the proof.

Corollary 1.13. Let $f \in C^{\infty}(a,b)$. Assume that there exist L, $M > 0$ such that, for any $n \in \mathbb{N} \cup \{0\}$ and for any $x \in I$:

$$\left| f^{(n)}(x) \right| \leq M L^n.$$

Then, for any $x_0 \in I$, $f(x)$ veryfies (1.11)

Definition. A function $f(x)$, defined on an open interval I, is analytic at $x_0 \in I$, if its Taylor series in x_0 converges to $f(x)$ in some neighbourhood of x_0.

The series expansion of the most important, commonly used functions can be inferred from equation (1.10), i.e. from the Taylor–Lagrange theorem.

Theorem 1.14. For any $x \in \mathbb{R}$, the following power series expansion holds:

- $e^x = \displaystyle\sum_{n=0}^{\infty} \frac{x^n}{n!}$
- $\sin x = \displaystyle\sum_{n=0}^{\infty} \frac{(-1)^n x^{2n+1}}{(2n+1)!}$
- $\cos x = \displaystyle\sum_{n=0}^{\infty} \frac{(-1)^n x^{2n}}{(2n)!}$
- $\sinh x = \displaystyle\sum_{n=0}^{\infty} \frac{x^{2n+1}}{(2n+1)!}$
- $\cosh x = \displaystyle\sum_{n=0}^{\infty} \frac{x^{2n}}{(2n)!}$

Theorem 1.15. The following power series converge for $|x| < 1$.

- $\dfrac{1}{1-x} = \sum_{n=0}^{\infty} x^n$
- $\dfrac{1}{(1-x)^2} = \sum_{n=0}^{\infty} (n+1)x^n$
- $-\ln(1-x) = \sum_{n=0}^{\infty} \dfrac{x^{n+1}}{n+1}$
- $\ln(1+x) = \sum_{n=0}^{\infty} (-1)^n \dfrac{x^{n+1}}{n+1}$
- $\arctan x = \sum_{n=0}^{\infty} (-1)^n \dfrac{x^{2n+1}}{2n+1}$
- $\operatorname{arctanh} x = \sum_{n=0}^{\infty} \dfrac{x^{2n+1}}{2n+1}$
- $(1+x)^\alpha = \sum_{n=0}^{\infty} \binom{\alpha}{n} x^n$
- $\arcsin x = \sum_{n=0}^{\infty} \dfrac{1}{2^{2n}} \binom{2n}{n} \dfrac{x^{2n+1}}{2n+1}$
- $\arcsin^2 x = \dfrac{1}{2} \sum_{n=1}^{\infty} \dfrac{1}{n^2 \binom{2n}{n}} (2x)^{2n}$
- $\operatorname{arcsinh} x = \sum_{n=0}^{\infty} \dfrac{(-1)^n}{2^{2n}} \binom{2n}{n} \dfrac{x^{2n+1}}{2n+1}$
- $\operatorname{arcsinh}^2 x = \dfrac{1}{2} \sum_{n=1}^{\infty} \dfrac{(-1)^{n-1}}{n^2 \binom{2n}{n}} (2x)^{2n}$

Power series representations are widely used in complex analysis, allowing us to extend the validity of Theorems 1.14 and 1.15 to the complex plane.

1.5 Complex analysis

We present a quick synopsis of some basic notions of complex analysis, which will be employed in the following chapters. For completeness, we introduce $\mathbb{C}$, the set of complex numbers.

Definition. $\mathbb{C}$ denotes the set of complex numbers. It is formed by the (ordered) couples (x, y) of real numbers, defining the sum and the multiplication as follows:

$$(x, y) + (u, v) = (x + u, y + v) \qquad (x, y) \cdot (u, v) = (xu - yv, xv + yu)$$

$\mathbb{C}$ is a field. The inverse of $(a, b) \neq (0, 0)$ is given by

$$\left(\frac{a}{a^2 + b^2}, \frac{-b}{a^2 + b^2} \right).$$

Notice that $\mathbb{R}$ is the sub-field of $\mathbb{C}$ whose elements are $(x, 0)$. Since

$$(x, y) = (x, 0) + (0, 1) \cdot (y, 0)$$

the complex number (x, y) can be represented as $z = x + i\,y$, being $i = (0, 1)$, where we write $x = (x, 0)$, $y = (y, 0)$. We can perform calculations with complex numbers using the new algebraic rules $i^2 = (0, 1) \cdot (0, 1) = (-1, 0) = -1$. If $z = x + i\,y \in \mathbb{C}$, then the real numbers x, y are called, respectively, real part and imaginary part of z; we use the notations Rez and Imz. The modulus of $z = x + i\,y$ is defined by $|z| := \sqrt{x^2 + y^2}$. The conjugate of $z \in \mathbb{C}$, say $\bar{z}$, is defined as $\bar{z} = x - i\,y$. We list the main properties of the conjugate operator, here z, z_1, z_2 are given complex numbers:

1. $\bar{\bar{z}} = z$,
2. $\overline{z_1 + z_2} = \bar{z}_1 + \bar{z}_2$,
3. $\bar{z} = z \iff z \in \mathbb{R}$,
4. $|z|^2 = z\bar{z}$,
5. $|\bar{z}| = |z|$,
6. $z + \bar{z} = 2\,\mathrm{Re}\ z$,
7. $\overline{z_1 z_2} = \bar{z}_1 \cdot \bar{z}_2$,
8. $z - \bar{z} = 2\,\mathrm{Im}\ z$.

The distance in $\mathbb{C}$ is defined as follows. For $z_1, z_2, \in \mathbb{C}$: $\mathrm{d}(z_1, z_2) = |z_1 - z_2|$. This Euclidean distance induces the topology in $\mathbb{C}$ exactly as in the case of $\mathbb{R}$ due to the following inequalities, known as triangular inequalities:

$$|z_1 + z_2| \leq |z_1| + |z_2|, \quad \big||z_1| - |z_2|\big| \leq |z_1 - z_2|.$$

A complex number $z \in \mathbb{C}$ is a point in the Cartesian plane: it can also be represented in polar form by $x = \rho\,\cos\varphi$, $y = \rho\,\sin\varphi$, where ρ is the modulus and φ the argument of the point $z = x + i\,y \in \mathbb{C}$, i.e. the angle formed by the vector ray joining the origin with the point z, where it is agreed to consider positive the anti-clockwise direction. The argument is, therefore, defined by the two relations:

$$\begin{cases} \cos\varphi = \dfrac{x}{\sqrt{x^2 + y^2}}, \\ \sin\varphi = \dfrac{y}{\sqrt{x^2 + y^2}}. \end{cases}$$

It follows that each complex number has infinite determinations of the argument. We call the principal argument the unique φ such that:

$$-\pi < \varphi \leq \pi.$$

De Moivre's formulæ

The polar representation makes it possible to express products, quotients and powers in a very meaningful way and, above all, allows the introduction of some remarkable representation formulae due to de Moivre. If:

$$z_1 = \rho_1 \cos\varphi_1 + i\rho_1 \sin\varphi_1, \quad z_2 = \rho_2 \cos\varphi_2 + i\rho_2 \sin\varphi_2,$$

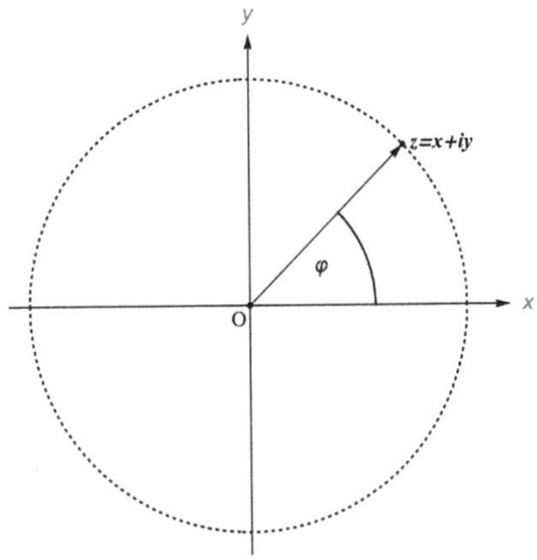

Figure 1.1: Polar form of a complex number

then, from the addition formulae of sine and cosine, it follows:

$$z_1 z_2 = \rho_1 \rho_2 \Big(\cos(\varphi_1 + \varphi_2) + i \sin(\varphi_1 + \varphi_2) \Big) .$$

This formula is known as de Moivre's first formula. If we take $n \in \mathbb{N}$, it makes sense to consider the powers $z \in \mathbb{C}$, and thus, the following particular case of the first formula is obtained:

$$z^n = \rho^n \left(\cos(n\,\varphi) + i \sin(n\,\varphi) \right) . \tag{1.12}$$

The formula (1.12) leads us to consider the roots of a complex number: given $n \in \mathbb{N}$, we say that $w \in \mathbb{C}$ is a n-th root of $z \in \mathbb{C}$ if $w^n = z$. If $z = \rho \cos \varphi + i \rho \sin \varphi$ the distinct n-simple roots of z are given by:

$$w = \sqrt[n]{\rho} \left(\cos \frac{\varphi + 2k\pi}{n} + i \sin \frac{\varphi + 2k\pi}{n} \right) , \quad k = 0, 1, \ldots, n-1 . \tag{1.13}$$

Equation (1.13) is de Moivre's second formula; note that the n-th root symbol here means the positive arithmetic root of the positive number ρ. Notice that n-th roots are non-defined for $z = 0$.

Functions of a complex variable

In the following, we will call *domain* an open and connected subset of $\mathbb{C}$.

Definition. Let $D \subseteq \mathbb{C}$ be a domain, $f : D \to \mathbb{C}$. The derivative of f at the point $z_0 \in D$ is defined by

$$f'(z_0) = \lim_{z \to z_0} \frac{f(z) - f(z_0)}{z - z_0} . \tag{1.14}$$

If the limit (1.14) exists for every $z_0 \in D$, $f(z)$, it is said to be holomorphic in D. When f is holomorphic in each $z \in \mathbb{C}$, we say that f is an entire function.

Remark. The concept of derivative has the same structure for functions of real variables as for functions of complex variables. On the one hand, this means that the theorems concerning the derivation of sums, products, and quotients of functions are immediately replicable. However, the limit (1.14) in the complex plane has more stringent implications because this limit is obtained isotropically with all possible directions in the complex plane. We will state, without proof, the properties that derive from the existence of the limit (1.14), referring to the literature mentioned earlier for the demonstrations of the following statements.

A function of a complex variable $f : \mathbb{C} \to \mathbb{C}$ can be represented as

$$f(z) = u(x, y) + i\, v(x, y)$$

having written $z \in \mathbb{C}$ in algebraic form $z = \text{Re}z + i\,\text{Im}z = x + i\,y$.

For example

$$f(z) = z^2 = (x + i\,y)^2 = x^2 - y^2 + i\,2xy.$$

The so-called Cauchy Riemann equations are the most important consequences of (1.14).

Theorem. If $f(z)$ is holomorphic, then the following identities hold:

$$\frac{\partial u}{\partial x} = -\frac{\partial v}{\partial y}, \quad \frac{\partial u}{\partial y} = -\frac{\partial v}{\partial x}.$$

It is immediate to verify that $f(z) = z^2 = (x + i\,y)^2 = x^2 - y^2 + i\,2xy$ is holomorphic while $f(z) = |z|^2 = x^2 + y^2$ is not holomorphic, for $z \neq 0$.

Even for complex cases, functions with the null derivative are constant under appropriate hypotheses.

Theorem. Given $r > 0$ and define the open disk

$$B(z_0, r) = \{z \in \mathbb{C} : |z - z_0| < r\}.$$

Assume $f : B(z_0, r) \to \mathbb{C}$ holomorphic and such that for any $z \in B(z_0, r)$, its derivative is zero, i.e. $f'(z) = 0$. Then there exists $w \in \mathbb{C}$ such that $f(z) = w$ for any $z \in B(z_0, r)$.

Theorem. If $f(z)$ is holomorphic in $z_0 \in D$, then there exists a complex sequence $(a_n) \subset \mathbb{C}$ such that:

$$f(z) = \sum_{n=0}^{\infty} a_n (z - z_0)^n. \tag{1.15}$$

Remark. In Section 1.2, we presented the theory of the sequences and series of functions; here, we remark that the existence of the limit (1.14) implies the fact that any holomorphic function can be represented with a convergent power series using (1.15). We have expressed in the Theorems 1.14 and 1.15 the main elementary functions in terms of serious powers. In light of what we have now seen, we are, therefore, dealing with holomorphic functions.

Note that from the power series representation of Theorem 1.14, follow some famous relations due to Euler

$$e^{i\,z} = \cos z + i\,\sin z \tag{1.16a}$$

$$\cos z = \frac{e^{i\,z} + e^{-i\,z}}{2} \tag{1.16b}$$

$$\sin z = \frac{e^{i\,z} - e^{-i\,z}}{2\,i} \tag{1.16c}$$

$$\sin(i\,z) = i\,\sinh z \tag{1.16d}$$

$$\cos(i\,z) = \cosh z. \tag{1.16e}$$

Definition. We will call a meromorphic function a holomorphic function in $z \in \mathbb{C}$ except at a finite number of points.

$f(z) = \dfrac{z}{z^2+1}$ is meromorphic, as it is holomorphic for $z \neq \pm\, i$.

Definition. Given a function $f : D \to \mathbb{C}$ and given $z_0 \in D$ being, as usual, D a domain in $\mathbb{C}$, we say that z_0 is a regular point if f is holomorphic in the disk $B(z_0, r)$ for some $r > 0$. If it is not the case, the point z_0 is called a singularity of f. If z_0 is a singularity of f and f is holomorphic at the punctured disc $B'(z_0, r) = B(z_0, r)\backslash\{z_0\}$ then z_0 is said an isolated singularity. The remaining singularities are called non-isolated or essential singularities.

In the previous example, $z_0 = i$ and $z_0 = -i$ are isolated singularities.

Integrals along curves

Definition. The function $\gamma : [a,\, b] \to \mathbb{C}$ is said to be a curve in $\mathbb{C}$. For our purposes we will assume $\gamma \in \mathcal{C}^1([a, b])$. We speak of a simple curve if for every $t,\, s \in [a,\, b]$, $t \neq s$ is true $\gamma(t) \neq \gamma(s)$. A simple closed curve is said closed if for each $t,\, s \in (a,\, b)$, $t \neq s$ we have $\gamma(t) \neq \gamma(s)$ while $\gamma(a) = \gamma(b)$.

The complex line integral of $f : D \to \mathbb{C}$ is defined when the simple curve verifies $\gamma(t) \in D$ and is given by:

$$\int_\gamma f(z)\,\mathrm{d}z = \int_a^b f(\gamma(t))\,\gamma'(t)\,\mathrm{d}t\,.$$

If γ is closed, it is customary to introduce the symbol of loop integral:

$$\oint_\gamma f(z)\,\mathrm{d}z\,.$$

For example, if $\gamma : [0,\, 2\pi] \to \mathbb{C}, \gamma(t) = \cos t + i\,\sin t,\; f(z) = \dfrac{1}{z}$ then

$$\oint_\gamma f(z)\,\mathrm{d}z = \int_0^{2\pi} \frac{(-\sin t + i\,\cos t)}{\cos t + i\,\sin t}\,\mathrm{d}t = \int_0^{2\pi} i\,\mathrm{d}t = 2\,\pi\,i\,.$$

While, keeping the same path γ, we consider $f(z) = z$ we find

$$\begin{aligned}\oint_\gamma f(z)\,\mathrm{d}z &= \int_0^{2\pi} (\cos t + i\,\sin t)\,(-\sin t + i\,\cos t)\,\mathrm{d}t \\ &= \int_0^{2\pi} \left(-2\sin t\cos t + i\,\left(\cos^2 t - \sin^2 t\right)\right)\,\mathrm{d}t = 0.\end{aligned}$$

The next couple of statements are of fundamental importance; they are due to Cauchy and Morera, respectively, and they prelude one of the most important results of complex analysis, known as the Cauchy integral formula.

Theorem. Let D denotes a domain in $\mathbb{C}$. Assume $f : D \to \mathbb{C}$ holomorphic and $\gamma : [a,b] \to \mathbb{C}$ simple closed curve $\gamma \subset D$. Then

$$\oint_\gamma f(z)\,\mathrm{d}z = 0\,.$$

The converse of Cauchy's theorem is due to Morera.

Theorem. If $f : D \to \mathbb{C}$, where D is a domain in $\mathbb{C}$, is continuous, such that for every simple closed curve $\gamma \subseteq D$ holds

$$\oint_\gamma f(z)\,\mathrm{d}z = 0,$$

then $f(z)$ is holomorphic.

We can state the fundamental result due to Cauchy.

Theorem (Cauchy integral formula). If $f : D \to \mathbb{C}$ is holomorphic in the domain D, and $\gamma \subset D$ is a simple closed curve if $z_0 \in \mathbb{C}$ lies inside γ then

$$\oint_\gamma \frac{f(z)}{z - z_0}\,\mathrm{d}z = 2\pi\,i\,f(z_0). \tag{1.17}$$

Observe that, from (1.17) follows:

$$\begin{aligned}f(z_0) &= \frac{1}{2\,\pi\,i} \oint_\gamma \frac{f(z)}{z - z_0}\,\mathrm{d}z \\ f'(z_0) &= \frac{1}{2\,\pi\,i} \oint_\gamma \frac{f(z)}{(z - z_0)^2}\,\mathrm{d}z \\ f^{(2)}(z_0) &= \frac{2}{2\,\pi\,i} \oint_\gamma \frac{f(z)}{(z - z_0)^3}\,\mathrm{d}z \\ &\cdots\cdots\cdots \\ f^{(n)}(z_0) &= \frac{n!}{2\,\pi\,i} \oint_\gamma \frac{f(z)}{(z - z_0)^{n+1}}\,\mathrm{d}z\end{aligned}$$

Analytic continuation

Definition 1.16. If $f : D \to \mathbb{C}$ is holomorphic on the domain D we say that $f_1 : D_1 \to \mathbb{C}$ is an analytic continuation of f if

- D_1 is a domain and $D \subset D_1$,
- $f_1 : D_1 \to \mathbb{C}$ is holomorphic,
- for any $x \in D$, $f(x) = f_1(x)$.

Roughly speaking, we are in the presence of an analytical continuation of a holomorphic function when its domain can be extended, and a holomorphic function defined in the greater domain, which coincides with the given function on the domain D, can be determined. The representation of an analytic function in a power series implies that the analytic prolongation of a holomorphic function is unique. This fact, which is also reasonably intuitive, will often be used in discussing special functions.

Laurent series and residues

Fixed $r > 0$ consider the pointed disk

$$B'(z_0, r) = \{z \in \mathbb{C} : 0 < |z - z_0| < r\}.$$

Let $f : B(z_0, r) \to \mathbb{C}$, holomorphic except in $z = z_0$. In this case, it is possible to show that $f(z)$ can be represented into a bilateral series, called Laurent series of the form

$$f(z) = \sum_{n=-\infty}^{+\infty} a_n (z - z_0)^n. \tag{1.18}$$

Using the Laurent series, we can provide a singularities taxonomy. Rewrite (1.18) as

$$f(z) = \sum_{n=-\infty}^{-1} a_n (z - z_0)^n + \sum_{n=0}^{+\infty} a_n (z - z_0)^n . \tag{1.19}$$

Observe that the second term on the right-hand side of (1.19) is holomorphic in a neighbourhood of z_0: the singularities are generated by the first term on the right-hand side of (1.19), which for this reason is called the principal part of the Laurent expansion.

Definition. The point z_0 is said

- a removable singularity if $a_n = 0$ for any $n < 0$;
- a pole of order $m \geq 1$ if $a_{-m} \neq 0$ and $a_n = 0$ for any $n < -m$;
- an isolated essential singularity if there is no m such that $a_n = 0$ for any $n < -m$.

Poles of order $1, 2, 3, \dots$ are said, simple, double, triple,...

We will then say that the pole is **manifest** if

$$f(z) = g(z)\,(z - z_0)^{-m} \tag{1.20}$$

with g holomorphic in $B(z_0, r)$ for some $r > 0$ and $g(z_0) \neq 0$, while in the other cases, we speak of a **hidden pole**.

For example, $z_0 = 0$ is a removable singularity for

$$f(z) = \frac{\sin z}{z}.$$

Expanding the sine in power series, see Theorem 1.14, we obtain

$$f(z) = \frac{\sin z}{z} = \frac{1}{z}\left(z - \frac{z^3}{3!} + \frac{z^5}{5!} + \dots\right) = 1 - \frac{z^2}{3!} + \frac{z^4}{5!} + \dots.$$

A similar argument shows that if $0 < |z| < +\infty$

$$f(z) = \frac{\sin z}{z^2}$$

has a simple (hidden) pole in $z_0 = 0$, in fact:

$$f(z) = \frac{\sin z}{z^2} = \frac{1}{z^2}\left(z - \frac{z^3}{3!} + \frac{z^5}{5!} + \dots\right) = \frac{1}{z} - \frac{z}{3!} + \frac{z^3}{5!} + \dots.$$

Finally for $0 < |z| < +\infty$ consider, recalling again Theorem 1.14:

$$f(z) = \sinh\frac{1}{z} = \sum_{n=0}^{\infty} \frac{z^{-(2n+1)}}{(2n+1)!}$$

We acknowledge that $z_0 = 0$ is an isolated essential singularity.

From a practical point of view, the following result makes it easy to determine the nature of polar singularities in many cases.

Theorem. If $f : B'(z_0, r) \to \mathbb{C}$ is holomorphic, f has a pole of order $m \in \mathbb{N}$ at z_0 if and only if

$$\lim_{z \to z_0} (z - z_0)^m \, f(z) = a \in \mathbb{C} \setminus \{0\}. \tag{1.21}$$

The formula (1.21) implies that the poles are naturally linked to the zeros of the reciprocal function.

Corollary. Suppose $f : B(z_0, r) \to \mathbb{C}$ holomorphic. Then f has a zero of order $m \in \mathbb{N}$ in z_0 if and only if $1/f$ has a pole of order m in z_0.

We observe that if f has in z_0 a pole of order m, the only negative index terms of the Laurent development other than zero are $a_{-1}, a_{-2}, \dots, a_{-m}$. Moreover, if $f(z)$ has in z_0 a pole of order m then $(z - z_0)^m f(z)$ is holomorphic.

Theorem. If $f : B'(z_0, r) \to \mathbb{C}$ is holomorphic and $\gamma \subset B(z_0, r)$ is a simple closed curve with $z_0 \in \mathbb{C}$ lies inside γ and assume that f has a pole of order m in z_0, hence the Laurent expansion of f is given by

$$f(z) = \sum_{n=-m}^{+\infty} a_n (z - z_0)^n .$$

The number:

$$a_{-1} = \frac{1}{2\pi\, i} \oint_\gamma f(z)\, \mathrm{d}z\,,$$

is said the **residue** of f in z_0 and is denoted by:

$$\operatorname{Res}(f, z_0).$$

Let us look at some practical guidelines for calculating residues, specifying the form of the function under study for the most frequently used situations. Given a function of the form:

$$f(z) = \frac{n(z)}{d(z)}$$

we see that f has a pole of order $m \in \mathbb{N}$ in $z_0 \in \mathbb{C}$ if there exists $r > 0$ such that

(i) $n(z)$, $d(z)$ are holomorphic in $B(z_0, r) := \{z \in \mathbb{C} : |z - z_0| < r\}$

(ii) $n(z_0) \neq 0$

(iii) $d(z)$ has a zero of order m in z_0 i.e.:

   (a) $d(z_0) = d'(z_0) = \cdots = d^{(m-1)}(z_0) = 0$, $d^{(m)}(z_0) \neq 0$

   (b) $d(z) = (z - z_0)^m\, g(z)$ with $g(z)$ holomorphic and $g(z_0) \neq 0$

Remark (Calculation of residues in simple poles)**.** We observe that if z_0 is a simple pole of $f : B'(z_0, r) \to \mathbb{C}$ we have

$$\operatorname{Res}(f, z_0) = \lim_{z \to z_0} (z - z_0)\, f(z).$$

If the simple pole is hidden, then

$$f(z) = \frac{n(z)}{d(z)} \implies \operatorname{Res}(f, z_0) = \frac{n(z_0)}{d'(z_0)}\,.$$

For example, if $a \in \mathbb{C}$ the function $f(z) = \dfrac{1}{z^2 + a^2}$ has hidden simple poles the $\pm\, i\, a$ and

$$\operatorname{Res}(f, \pm\, i\, a) = \pm \frac{1}{2\, i\, a} = \mp \frac{i}{2a}\,.$$

Remark (Calculation of residues in multiple poles). If f has a manifest pole of order m in $z_0 \in \mathbb{C}$ then, recalling (1.20):

$$\operatorname{Res}(f, z_0) = \frac{1}{(m-1)!}\, g^{(m-1)}(z_0).$$

In the case of a hidden pole of order m, it is convenient to manipulate the expression of $f(z)$ in such a way that it corresponds to the case of a manifest pole: e.g. if we want the residue in $i\,a$ (double pole) of

$$f(z) = \frac{1}{(z^2+a^2)^2} = \frac{1}{(z+i\,a)^2}\,\frac{1}{(z-i\,a)^2} = g(z)\,\frac{1}{(z-i\,a)^2}$$

where $g(z) = \dfrac{1}{(z+i\,a)^2}$, whence $g'(i\,a) = -\dfrac{i}{4a^3}$ and, therefore

$$\operatorname{Res}(f, i\,a) = -\frac{i}{4a^3}.$$

We state the so-called residue theorem.

Theorem 1.17. If $D \subseteq \mathbb{C}$ is a domain, $\gamma \subseteq D$ is a simple closed curve and f a holomorphic function in γ and within it except in m isolated singularities $a_1, \ldots, a_m$. Then:

$$\oint_\gamma f(z)\,\mathrm{d}z = 2\pi\,i \sum_{h=1}^{m} \operatorname{Res}(f, a_h).$$

The residue theorem greatly impacts the calculation of real integrals. Let's examine some interesting illustrative examples.

Example. We use Theorem 1.17 to evaluate the (real) integral

$$\int_0^{+\infty} \frac{\mathrm{d}x}{(4+x^2)\,(1+x^2)^2}\,.$$

We integrate

$$f(z) = \frac{1}{(4+z^2)\,(1+z^2)^2}$$

along the simple closed curve $\gamma = \Gamma(0,r) \cup [-r, r]$ shown in the Figure 1.2.

The curve γ, which consists of the semicircle with centre in the origin and radius R lying in the first and second quadrant and diameter, located on the x-axis, is traversed anti-clockwise, as indicated by the arrow in Figure 1.2. Taking care to choose $r > 2$, f is holomorphic on γ and within it except for a double pole in i and a simple pole in $2\,i$. We can invoke the residue Theorem 1.17 to obtain

$$\oint_\gamma f(z)\,\mathrm{d}z = \int_{-r}^{r} f(z)\,\mathrm{d}z + \int_{\Gamma(0,r)} f(z)\,\mathrm{d}z = 2\pi\,i\,(\operatorname{Res}(f,i) + \operatorname{Res}(f,2i))$$

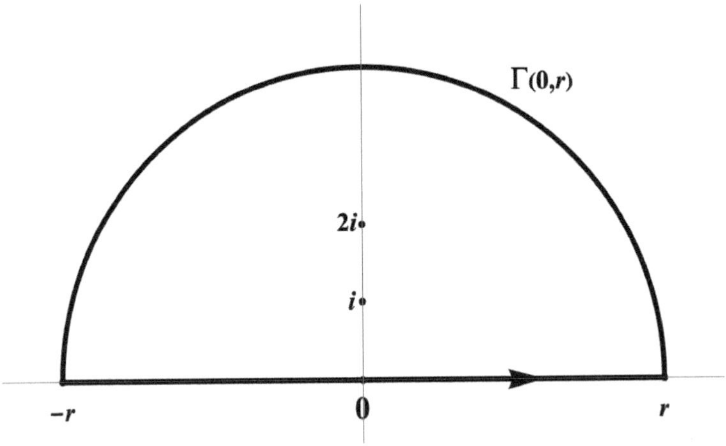

Figure 1.2: $\gamma = \Gamma(0,r) \cup [-r,r]$

But since we have

$$\mathrm{Res}\,(f,i) = -\frac{i}{36}, \quad \mathrm{Res}\,(f,2i) = -\frac{i}{36}$$

we infer

$$\int_{-R}^{R} f(z)\,\mathrm{d}z + \int_{\Gamma(0;R)} f(z)\,\mathrm{d}z = 2\pi\, i\left(-\frac{i}{36} - \frac{i}{36}\right) = \frac{\pi}{9}.$$

Observed then that, the general inequality holds, being $\ell(\gamma)$ the length of the closed curve γ

$$\left|\int_{\gamma} f(z)\,\mathrm{d}z\right| \leq \int_{a}^{b} |f(\gamma(t))\gamma'(t)|\,\mathrm{d}t \leq \sup_{z\in\gamma} |f(z)|\,\ell(\gamma)$$

we see that

$$\left|\int_{\Gamma(0,r)} f(z)\,\mathrm{d}z\right| \to 0 \text{ as } r \to +\infty$$

while

$$\int_{-r}^{r} f(z)\,\mathrm{d}z \to 2\int_{0}^{\infty} f(z)\,\mathrm{d}z \text{ as } r \to +\infty.$$

Thus, we can arrive at the desired integral

$$\int_{0}^{+\infty} \frac{\mathrm{d}x}{(4+x^2)\,(1+x^2)^2} = \frac{\pi}{18}.$$

We present a second example, which will be useful to obtain an important formula, page 72 in chapter 2.

Example. For $0 < a < 1$ we show that

$$\int_{-\infty}^{+\infty} \frac{e^{ax}}{1+e^{x}} \mathrm{d}x = \frac{2\,\pi\, i e^{a\,\pi\, i}}{e^{2a\pi\, i}-1} = \frac{\pi}{\sin(a\,\pi)}. \tag{1.22}$$

Let us consider the complex function $f(z) = e^{az}(1+e^{z})^{-1}$ on the rectangular contour in Figure 1.3 of vertices $(\pm r, \pm r + i\,2\pi)$ where $r > 0$. f has a simple

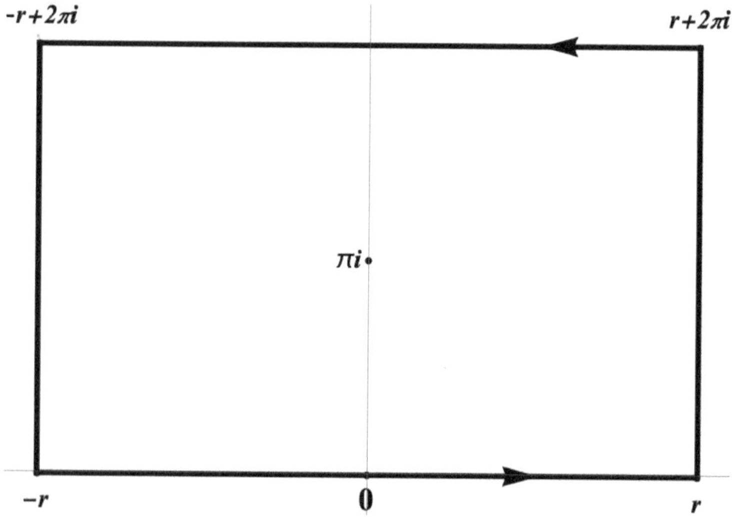

Figure 1.3: Rectangular contour

hidden pole in $z = i\,\pi$ hence:

$$f(z) = \frac{n(z)}{d(z)} \implies \mathrm{Res}\,(f, a) = \frac{n(a)}{d'(a)} \implies \mathrm{Res}\,(f, i\,\pi) = \frac{e^{a\,i\,\pi}}{e^{i\,\pi}} = -e^{a\,i\,\pi}.$$

From the residue Theorem 1.17, we infer

$$-\,2\pi\, i\, e^{a\pi\, i} = \int_{-r}^{r} \frac{e^{ax}}{1+e^{x}} \mathrm{d}x + \int_{0}^{2\pi} \frac{e^{a(r+i\,y)}}{1+e^{r+i\,y}}\, i\,\mathrm{d}y$$

$$-\int_{-r}^{r} \frac{e^{a(x+2\pi\, i)}}{1+e^{a(x+2\pi\, i)}} \mathrm{d}x - \int_{0}^{2\pi} \frac{e^{a(-r+i\,y)}}{1+e^{-r+i\,y}}\, i\,\mathrm{d}y.$$

Observed that if r is sufficiently large, the following applies:

$$\left|\frac{e^{a(r+i\,y)}}{1+e^{r+i\,y}}\right| \le \frac{e^{ar}}{e^{r}-1} \le 2e^{-(1-a)\,r}$$

$$\left|\frac{e^{a(-r+i\,y)}}{1+e^{-r+i\,y}}\right| \le \frac{e^{-a\,r}}{1-e^{-r}} \le 2e^{-a\,r}$$

$$\int_{-r}^{r} \frac{e^{a(x+2\pi\, i)}}{1+e^{a(x+2\pi\, i)}}\, \mathrm{d}x = e^{2a\pi\, i} \int_{-r}^{r} \frac{e^{ax}}{1+e^{ax}}\, \mathrm{d}x$$

taking the limit as $r \to +\infty$ we arrive at

$$\left(1 - e^{2a\pi\, i}\right) \int_{-\infty}^{+\infty} \frac{e^{ax}}{1+e^{x}} \mathrm{d}x = -2\,\pi\, i e^{a\,\pi\, i}\,,$$

which proves (1.22); notice that the last step follows from (1.16c).

1.6 Differentiation under integral sign

Calculating integrals using exact formulæ is invaluable in all situations where integrals come into play, and the field of Special Functions is undoubtedly no exception. In many cases, the elementary techniques do not allow the evaluation of definite integrals, which nevertheless have great importance. The most celebrated is the so-called Probability integral.

$$\int_{-\infty}^{\infty} e^{-\pi\, x^2} \mathrm{d}x = 1\,.$$

Differentiation under the integral sign is a technique that allows the computation of families of integrals depending on one or more parameters. This technique requires some hypotheses, which will be set out rigorously in this section. To give an idea of how to proceed in practice, let's give an example of an application without, at least in this preliminary phase, worrying about the problem of the correctness of the passages, which will be specified later.

Example. To illustrate the technique, we show that

$$\int_{0}^{+\infty} \frac{\sin t}{t}\, \mathrm{d}t = \frac{\pi}{2}\,.$$

This integral, which we will also calculate using Eulerian integrals in Section 2.11, is known as the Dirichlet integral. Let us consider the function which represents a family of integrals depending on the parameter x.

$$F(x) = \int_{0}^{+\infty} e^{-xt}\, \frac{\sin t}{t}\, \mathrm{d}t, \quad x > 0\,.$$

The integral we are trying to calculate is indeed $F(0)$ or, more precisely, $\lim\limits_{x\to 0+} F(x)$, since we do not have a priori knowledge concerning the continuity of F as a function of x. Now, assume that we can exchange between derivation and integration (indeed, this is nothing else that exchange between a limit and an integral), obtaining

$$F'(x) = \frac{\mathrm{d}}{\mathrm{d}x}\left(\int_0^{+\infty} e^{-xt}\,\frac{\sin t}{t}\,\mathrm{d}t\right) = \int_0^{+\infty} \frac{\mathrm{d}}{\mathrm{d}x}\left(e^{-xt}\,\frac{\sin t}{t}\right)\,\mathrm{d}t.$$

In conclusion:

$$F'(x) = -\int_0^{+\infty} e^{-xt}\sin t\,\mathrm{d}t = -\frac{1}{1+x^2}.$$

In this way, even if the problem of determining $F(x)$ cannot be faced with elementary methods, we know its derivative precisely; therefore, we need an integration which, in this case, is fortunately immediate for us, to conclude that, using a constant of integration $c \in \mathbb{R}$ that $F(x) = c - \arctan x$. The constant c is obtained by observing that:

$$\lim_{x\to+\infty} F(x) = c - \frac{\pi}{2}.$$

While referring to the definition, we find

$$|F(x)| = \left|\int_0^{+\infty} e^{-xt}\,\frac{\sin t}{t}\,\mathrm{d}t\right| \leq \int_0^{+\infty} e^{-xt}\,\mathrm{d}t = \frac{1}{x}.$$

Hence

$$\lim_{x\to+\infty} F(x) = 0\,.$$

This determines the value of the integration constant $c \in \mathbb{R}$ and the expression of $F(x)$

$$F(x) = \frac{\pi}{2} - \arctan x$$

So, if we assume that F is continuous for $x = 0$ and we have

$$\frac{\pi}{2} = F(0) = \int_0^{+\infty} \frac{\sin t}{t}\,\mathrm{d}t.$$

This example illustrates the effectiveness of this method: there are many applications to the calculation of integrals, and whoever is interested can consult [85] chapter 3; our aim here is not so much to recall the multiple applications of the method, in consideration of the fact that these will naturally occur in our future discussion, but instead to state under what conditions it is possible to differentiate under the sign of integral. We begin with the case characterised by the most restrictive hypotheses, which does not apply to the example just treated.

Theorem 1.18. Given real numbers $a < b$ and $c < d$ suppose $f : [a, b] \times [c, d] \to \mathbb{R}$ continuous. For $x \in [a, b]$ we define

$$F(x) = \int_c^d f(x,t)\,\mathrm{d}t\,.$$

If f it is partially differentiable with respect to x and if $\frac{\partial f}{\partial x} : [a,b] \times [c,d] \to \mathbb{R}$ is continuous then for any $a < x < b$ we have that $F(x)$ is differentiable and

$$F'(x) = \int_c^d \frac{\partial f}{\partial x}(x,t)\,\mathrm{d}t\,.$$

For the proof, see [94] chapter 8, Theorem 1. Here, we limit to say that, working in the compact rectangle $[a, b] \times [c, d]$, uniform continuity is also ensured (Heine's theorem), which is the main ingredient of the proof.

Theorem 1.18, as we have just stated does not apply to the previous example, as in that case, the integral is defined in an unbounded interval.

We must add a condition to ensure the possibility of differentiating under the integral sign and the problem in which the integrand presents singularities at some point in the interval. We start with the case of finite singularity.

Theorem 1.19. Given real numbers $a < b$ and $c < d$ suppose $f : [a,b] \times [c,d) \to \mathbb{R}$ continuous. Moreover, assume that there exists $g : [c,d) \to \mathbb{R}$ such that for any $x \in [a,b]$ and $t \in [c,d)$:

$$|f(x,t)| \leq g(t) \quad \text{and} \quad \int_c^d g(t)\,\mathrm{d}t \quad \text{converges.}$$

The improper integral

$$F(x) = \int_c^d f(x,t)\,\mathrm{d}t$$

is well defined for any $x \in [a,b]$ and $F(x)$ is indeed continuous. Moreover, if we assume that $\frac{\partial f}{\partial x} : [a,b] \times [c,d) \to \mathbb{R}$ is continuous and that exists $h : [c,d) \to \mathbb{R}$ such that for any $x \in [a,b]$ and $t \in [c,d)$:

$$\left|\frac{\partial f}{\partial x}(x,t)\right| \leq h(t) \quad \text{and} \quad \int_c^d h(t)\,\mathrm{d}t \quad \text{converges, then}$$

$$F'(x) = \int_c^d \frac{\partial f}{\partial x}(x,t)\,\mathrm{d}t.$$

The same also holds in the unbounded case $d = +\infty$.

Remark. Returning to the Dirichlet integral, we observe that we have

$$\left|\frac{\partial f(x,t)}{\partial x}\right| = \left|-e^{-x\,t}\sin t\right| \leq e^{-x\,t} \leq \begin{cases} e^{-t} & \text{for } x \geq 1 \\ 1 & \text{for } 0 \leq x < 1. \end{cases}$$

The hypotheses of Theorem 1.19 are satisfied and authorised to differentiate the parametric integral, validating our previous calculation.

1.7 Fourier series

The theory of the Fourier series is fundamental in the study of Special Functions. The scientific literature dedicated to the Fourier series is immense, considering its numerous applications in varied contexts. Here, we limit ourselves to indicating the texts that inspired our very brief synopsis, which is aimed at the aspects we will use in the following: [21, 39, 44]. Fourier series were introduced due to the need to study oscillatory phenomena: the basic idea is to represent a periodic function as the sum of periodic functions, called elementary harmonics, obtained through more straightforward trigonometric functions. We fix the basic notations and terminology.

Definition. $f : \mathbb{R} \to \mathbb{R}$ is said periodic, with period $T > 0$ if $f(x+T) = f(x)$ for any $x \in \mathbb{R}$.

Observe that if f is periodic with period T then

i) f is also periodic with period mT for any $m \in \mathbb{N}$.

ii) In general it cannot be excluded that f is periodic with period T/m for some $m \in \mathbb{N}$.

iii) To determine the behaviour of f, it is sufficient to know its behaviour on any interval of length equal to its period T: usually, the interval $[0, T)$ or the interval $[-T/2, T/2)$ is chosen.

iv) If f is constant it is periodic with period T for any $T > 0$.

v) If f is non constant then $T_0 = \inf\{ T > 0 : f(x+T) = f(x) \}$ is said fundamental period of f.

vi) If f is integrable in $[0, T]$ then for any $a \in \mathbb{R}$, $\displaystyle\int_0^T f(x)\,\mathrm{d}x = \int_a^{T+a} f(x)\,\mathrm{d}x$.

vii) For any $S > 0$ then $g(x) = f(\frac{T}{S}\, x)$ is periodic: its period is S.

From this last property, we see that, as several authors have done, it is not restrictive to limit the case to periodic functions with a predetermined period; the easiest choice, given the use of elementary trigonometric functions, which are the most natural example of periodic functions, is to take $T = 2\,\pi$. We call trigonometric polynomial any finite linear combination of trigonometric functions of the form:

$$p_n(x) = a_0 + \sum_{k=1}^{n} \left(a_k \cos(k\,x) + b_k \sin(k\,x) \right),$$

where $a_0, a_1, \dots a_n, b_1, \dots b_n \in \mathbb{R}$. Fourier series theory aims to establish, under appropriate hypotheses, when an assigned 2π-periodic function of the real variable x, say $f(x)$, can be representable with a trigonometric polynomial, or when this is not possible, as the limit, in some sense, we are going to establish, of a sequence of trigonometric polynomials. Naturally, this implies that we must understand how, once the periodic function f has been assigned, to generate the sequences of coefficients (a_n) and (b_n) in an appropriate way to obtain a series of functions that converge to f itself. The existence of a sequence of functions converging to the function f is ensured by the Hilbert space theory. For an introduction to Hilbert spaces, we refer, for instance, to [116]. In any case, we do not need this generality; we need to identify an appropriate class of functions for which the approximation we will determine converges to the function that generated it.

Fourier coefficients

The sub-class of 2π-periodic functions that we will approximate is not, as it might seem natural, that of continuous functions, but the more extended one of the piecewise continuous functions. A piecewise continuous function is any function that in the interval $[0, 2\pi]$ has at most a finite number of points of discontinuity, which we will denote by $x_1, \ldots, x_m$ of the jump type, i.e. such that both limits:

$$\lim_{x \to x_k^-} f(x) := f(x_k^-), \qquad \lim_{x \to x_k^+} f(x) := f(x_k^+),$$

are finite for each of the discontinuity points $1 \leq k \leq m$. Moreover, we say that f is regularised if for any points of discontinuity x_k

$$f(x_k) = \frac{1}{2}\left(f(x_k^-) + f(x_k^+)\right). \tag{1.23}$$

We summarise these concepts in the following definition:

Definition 1.20. We indicate with $\mathcal{C}_r([0, 2\pi])$ the vector space of the 2π-periodic, piecewise continuous regularised functions.

Of course, we took the liberty of including the vector nature of $\mathcal{C}_r([0, 2\pi])$ in definition 1.20, which is easily demonstrated.

Definition. A real two-variable mapping $\langle \cdot \mid \cdot \rangle$ is called inner product it is

- symmetric, i.e. $\langle f \mid g \rangle = \langle g \mid f \rangle$
- linear, i.e. $\langle f + g \mid h \rangle = \langle f \mid h \rangle + \langle g \mid h \rangle$
- homogenous, i.e. $\langle \lambda f \mid g \rangle = \lambda \langle f \mid g \rangle$
- positive, i.e. $\langle f \mid f \rangle \geq 0, \qquad \langle f \mid f \rangle = 0 \iff f = 0$

Observe that the bilinear mapping

$$\langle f \mid g \rangle = \int_0^{2\pi} f(x)\, g(x)\, \mathrm{d}x$$

is an inner product. This inner product induces a metric in $\mathcal{C}_r\,([0,2\pi])$, with norm if $f \in \mathcal{C}_r\,([0,2\pi])$ given by

$$\|f\| = \left(\int_0^{2\pi} f^2(x)\, \mathrm{d}x \right)^{1/2}.$$

Observe that the numerable family of functions:

$$\frac{1}{\sqrt{2\pi}}, \frac{1}{\sqrt{\pi}} \cos x, \frac{1}{\sqrt{\pi}} \sin x, \ldots, \frac{1}{\sqrt{\pi}} \cos mx, \frac{1}{\sqrt{\pi}} \sin mx, \ldots$$

constitutes what is called an orthonormal system, in the sense that, having chosen two functions, say f, g in the system, we have that

$$\langle f \mid g \rangle = \begin{cases} 1 & \text{if } f = g, \\ 0 & \text{if } f \neq g. \end{cases}$$

This is due to the following identities, which depend on the computation of elementary trigonometric integrals, here $m,\, n \in \mathbb{N}$:

$$\begin{aligned} &\int_0^{2\pi} \cos(mx)\, \mathrm{d}x = \int_0^{2\pi} \sin(mx)\, \mathrm{d}x = \int_0^{2\pi} \cos(mx)\, \sin(nx)\mathrm{d}x = 0; \\ &\int_0^{2\pi} \cos(mx)\, \cos(nx)\, \mathrm{d}x = \int_0^{2\pi} \sin(mx)\, \sin(nx)\, \mathrm{d}x = 0, \\ &\int_0^{2\pi} \cos^2(mx)\, \mathrm{d}x = \int_0^{2\pi} \sin^2(mx)\, \mathrm{d}x = \pi. \end{aligned} \tag{1.24}$$

Definition. A Fourier series is a series of functions of the form

$$a_0 + \sum_{m=1}^{\infty} (a_m \cos(m\,x) + b_m \sin(m\,x)). \tag{1.25}$$

The real numbers a_0, a_m, b_m, $m \in \mathbb{N}$ are said Fourier coefficients.

Fourier series are the primary tool for representing periodic functions. Observe that if the series converges pointwise to a function f, this is necessarily periodic with fundamental period of 2π. Of course, the convergence depends on the Fourier coefficients. For example, if the series,

$$\sum_{m=1}^{\infty} a_m, \quad \sum_{m=1}^{\infty} b_m\,,$$

are absolutely convergent, it follows that the Fourier series (1.25) converges totally as a consequence of the straightforward inequality

$$|a_m \cos(mx) + b_m \sin(mx)| \leq |a_m| + |b_m| \,.$$

It follows that the function obtained by series summation of the form

$$f(x) = a_0 + \sum_{m=1}^{\infty} (a_m \cos(n\,x) + b_m \sin(m\,x)) \tag{1.26}$$

is continuous, and, more importantly, we can establish how coefficients a_m and b_m are linked to $f(x)$. Indeed, multiplying both sides of (1.26) by $\cos(nx)$ having chosen $n \in \mathbb{N} \cup \{0\}$ and integrating in $[0, 2\pi]$ we obtain

$$\begin{aligned}\int_0^{2\pi} f(x)\,\cos(nx)\,\mathrm{d}x = a_0 \int_0^{2\pi} \cos(nx)\,\mathrm{d}x \\ + \int_0^{2\pi} \cos(nx) \left(\sum_{m=1}^{\infty} (a_m \cos(m\,x) + b_m \sin(m\,x)) \right) \mathrm{d}x \,.\end{aligned}$$

Assuming that the Fourier series is totally convergent, we can exchange the integration and the summation and, ultimately, invoke the integral relations (1.24) so that we arrive at

$$\int_0^{2\pi} f(x)\,\cos(nx)\,\mathrm{d}x = \begin{cases} 2\pi a_0 & \text{if } n = 0 \\ \pi\,a_n & \text{if } n > 0 \end{cases}$$

Proceeding in the same way, this time multiplying by $\sin(nx)$ and then integrating in $[0, 2\pi]$ we see that

$$\int_0^{2\pi} f(x)\,\sin(nx)\,\mathrm{d}x = \pi\,b_n \,.$$

Therefore, we have thus calculated the Fourier coefficients:

$$\begin{aligned} a_0 &= \frac{1}{2\pi} \int_0^{2\pi} f(x)\,\mathrm{d}x, \\ a_n &= \frac{1}{\pi} \int_0^{2\pi} f(x)\,\cos(nx)\,\mathrm{d}x,\ n \in \mathbb{N}, \\ b_n &= \frac{1}{\pi} \int_0^{2\pi} f(x)\,\sin(nx)\,\mathrm{d}x,\ n \in \mathbb{N}. \end{aligned} \tag{1.27}$$

Even if it does not fall within the scope of our discussion, it is possible to verify that the relations (1.27) that define the Fourier coefficients are valid in the general contest of functions $\mathcal{C}_r\left([0, 2\pi]\right)$, however, the convergence of the series could be neither total nor pointwise, we will examine this in more detail later. The structure of the formulæ (1.27) that identifies the Fourier coefficients has significant consequences when applied to even or odd functions. For the sake of clarity, we state the following proposition:

Proposition 1.21. Let $f \in \mathcal{C}_r([0, 2\pi])$. If f is an even function, then:

$$a_0 = \frac{1}{\pi} \int_0^{\pi} f(x)\, \mathrm{d}x\,.$$

For any $n \in \mathbb{N}$

$$a_n = \frac{2}{\pi} \int_0^{\pi} f(x) \cos(nx)\, \mathrm{d}x, \qquad b_n = 0\,.$$

If f is an odd function, then for any $n \in \mathbb{N}$:

$$a_0 = 0, \quad a_n = 0, \quad b_n = \frac{2}{\pi} \int_0^{\pi} f(x) \sin(nx)\, \mathrm{d}x\,.$$

Proof. The proof depends on the direct computation of the integrals, remembering that since $f \in \mathcal{C}_r([0, 2\pi])$, then

$$\int_0^{2\pi} f(x) \cos(nx)\, \mathrm{d}x = \int_{-\pi}^{\pi} f(x) \cos(nx)\, \mathrm{d}x,$$
$$\int_0^{2\pi} f(x) \sin(nx)\, \mathrm{d}x = \int_{-\pi}^{\pi} f(x) \sin(nx)\, \mathrm{d}x.$$

Then, according to the parity of f, the appropriate symmetry properties are applied, from which the statement follows. □

Finally, we observe that it is possible to construct the Fourier series by determining its coefficients through the relations (1.27) also for non-regularised periodic functions. Indeed, the Fourier series thus obtained coincides with the function obtained by regularising the assigned function.

Quadratic convergence of the Fourier Series

The analysis of the convergence of the Fourier series is crucial for applications. Here, we will state several necessary conditions without going into the details of the proofs. Specifically, we will deal with pointwise convergence, uniform convergence obtained through total convergence, and quadratic convergence, which is naturally induced by the inner product but is less manageable in applications.

Consider the trigonometric polynomial, obtained by truncating series (1.25):

$$p_n(x) = a_0 + \sum_{k=1}^{n} \left(a_k \cos(k\,x) + b_k \sin(k\,x)\right).$$

If f is set by (1.25) we introduce the quadratic remainder between p_n and f

$$Q_n = \int_0^{2\pi} \left(f(x) - p_n(x)\right)^2 \mathrm{d}x.$$

It is worth mentioning that the Fourier coefficients are such that, among all the possible trigonometric polynomials, they minimise the quadratic deviation; that is, they possess the property of least squares. Moreover, using the relevant computations, the quadratic remainder is given by

$$Q_n = \int_0^{2\pi} f^2(x)\,\mathrm{d}x - 2\pi\, a_0^2 - \pi \sum_{k=1}^{n} (a_k^2 + b_k^2).$$

Theorem. Let $f \in \mathcal{C}_r\left([0, 2\pi]\right)$. The Fourier series generated by f converges to f in "quadratic norm", that is

$$\lim_{n\to+\infty} Q_n = 0.$$

Moreover, the identity, due to Parseval, holds;

$$\int_0^{2\pi} f^2(x)\,\mathrm{d}x = \pi \left(2\, a_0^2 + \sum_{n=1}^{\infty} (a_n^2 + b_n^2) \right). \tag{1.28}$$

An aspect of a specific technical nature, but indeed of interest in applications, consists of the periodic extension of a given function. Suppose we want to apply the results obtained on the Fourier series, such as the Parseval identity (1.28) just stated, to a function that by its nature is not periodic. In that case, we consider its restriction to the interval $[-\pi,\ \pi]$ and then replicate its behaviour periodically on the right-hand side from the intervals $[\pi,\ 3\pi]$, $[3\pi,\ 5\pi]\ldots$ and on the left-hand side on $[-3\pi,\ -\pi],\ldots$ this is expressed formally by the formula, where $\tilde{f}(x)$ denotes the periodic extension of f:

$$\tilde{f}(x) := f\left(x - 2\pi \left\lfloor \frac{\pi + x}{2\pi} \right\rfloor \right) \tag{1.29}$$

where the symbol $\lfloor\, x\, \rfloor$ denotes the integer part of x. The complication of the formula (1.29) need not worry; it is the rigorous definition for the extension, and this very formal representation is easy to implement with any symbolic calculus software. For instance, in the case of $f(x) = x$ in the interval $[-\pi,\ \pi]$, the plot of the periodic extension is shown in Figure 1.4 below.

Basel and related problems

Using the Fourier series, it is quite easy to sum up some very important series; first of all, the series which represents one of the most celebrated problems in Mathematics, which consists in determining the exact value of the

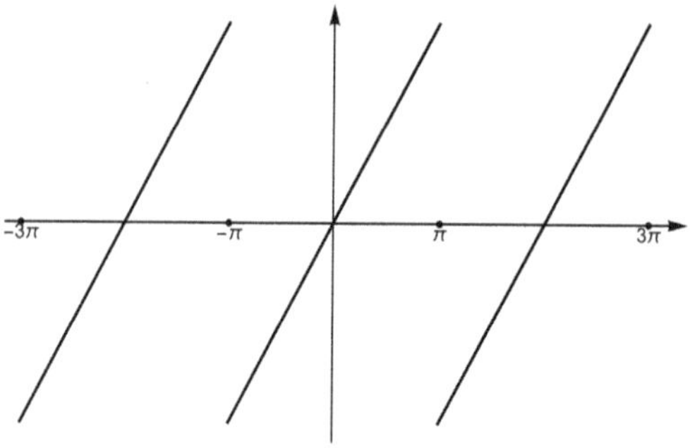

Figure 1.4: Plot of periodic extension $f(x) = x$

infinite series:

$$\sum_{n=1}^{\infty} \frac{1}{n^2} = \frac{\pi^2}{6}. \tag{1.30}$$

Pietro Mengoli (1626–1686) originally posed this problem in 1644, and Euler provided the first solution in [48]. The problem is called "Basel problem", referring to Euler's hometown. However, it was formulated in Bologna, where Mengoli taught. Anyway at the time, the city of Basel, due to the scientific activity carried out there by the brothers Jakob and Johann Bernoulli of Daniel Bernoulli, son of Johann and, of Euler himself, was, in fact, the pulsating centre of mathematical research. The solution proposed by Euler uses the representation of the sine by infinite product, which we will prove later. Using Fourier analysis, here, we solve the Basel problem using Parseval identity (1.28), applied to $f(x) = x$ in the interval $[-\pi, \pi]$, or more precisely to its periodic extension obtained using (1.29). Since this function is odd we can refer to Proposition 1.21 to infer that $a_n = 0$ for each integer $n \geq 0$ while for $n \in \mathbb{N}$:

$$b_n = \frac{2}{\pi} \int_0^{\pi} x \sin(nx)\, \mathrm{d}x = (-1)^{n+1} \frac{2}{n}.$$

Hence from (1.28) follows that

$$\int_{-\pi}^{\pi} x^2 \mathrm{d}x = \pi \sum_{n=1}^{\infty} b_n^2 = 4\pi \sum_{n=1}^{\infty} \frac{1}{n^2}$$

which, indeed, shows (1.30), after one has computed the elementary integral.

Using identity (1.28) again in the interval $[-\pi, \pi]$ but with $f(x) = |x|$ we can find the sum of a series related to (1.30), namely:

$$\sum_{n=1}^{\infty} \frac{1}{(2n-1)^4}.$$

In this case, working with an even function Proposition 1.21 leads to

$$a_0 = \frac{2}{\pi}\int_0^\pi |x|\,\mathrm{d}x = \frac{2}{\pi}\int_0^\pi x\,\mathrm{d}x = \pi,$$
$$a_n = \frac{2}{\pi}\int_0^\pi x\cos(nx)\,\mathrm{d}x = \frac{2(\cos(n\,\pi)-1)}{\pi\,n^2}.$$

Noting that

$$\cos(n\,\pi) - 1 = \begin{cases} -2 & \text{if } n \text{ is odd} \\ 0 & \text{if } n \text{ is even.} \end{cases}$$

We infer that, for $n \geq 1$

$$a_n^2 = \frac{16}{\pi^2(2n-1)^4}.$$

On the other side, since $\sin(n\,x)$ is an odd function, it is immediate to see that $b_n = 0$ for any $n \in \mathbb{N}$. Henceforth (1.28) implies

$$\frac{2}{3}\pi^2 = \frac{\pi^2}{2} + \sum_{n=1}^{\infty} \frac{16}{\pi^2(2n-1)^4}.$$

Thus, we arrive at

$$\sum_{n=1}^{\infty} \frac{1}{(2n-1)^4} = \left(\frac{2}{3} - \frac{1}{2}\right)\frac{\pi^4}{16} = \frac{\pi^4}{96}.$$

Pointwise and uniform convergence

The quadratic convergence of the Fourier series is a consequence of the norm induced by the inner product in the Hilbert space. Having an interest in applications, it is essential to understand if and how this convergence implies pointwise or uniform convergence. In general, there is no direct implication, but necessary conditions may be established; we will discuss some of them in this section, following the classical monograph [102] to which we refer for the proofs, which we will not exhibit here. The first of the two statements we present concerns pointwise convergence and is essentially the consequence of a Dini's Theorem, which we do not show here, for which we refer to [102].

Theorem 1.22. Consider $f \in \mathcal{C}_r([0, 2\pi])$. Let $x_0 \in [0, 2\pi]$ a finite jump discontinuity for f and this case there exists $\delta > 0$ such that the quotients

$$\frac{f(x_0 + x) - f(x_0^+)}{x} \quad \text{and} \quad \frac{f(x_0 - x) - f(x_0^-)}{x},$$

are bounded for $|x| < \delta$. Then the Fourier series for f converges at x_0 at the arithmetic means of the jump i.e. the value $f(x_0)$ introduced in (1.23). If x_0 is a point of continuity, the Fourier series converges at $f(x_0)$.

Remark. This result motivates the notion of a regularised function as the Fourier series converges to the average of the two values in the jump-type discontinuity points.

To achieve uniform convergence, we use a growth condition, see (1.31) below, known as Hölder continuity.

Theorem 1.23. Consider $f \in \mathcal{C}_r([0, 2\pi])$ assume that f is continuous and moreover that there exist $L, \alpha > 0$ such that for any $x, y \in [0, 2\pi]$

$$|f(x) - f(y)| \leq L\,|x - y|^{\alpha}\,. \tag{1.31}$$

Then the Fourier series for f converges uniformly and represents f.

Remark. For $\alpha = 1$ (1.31) is a Lipschitz continuity condition, i.e.:

$$|f(x) - f(y)| \leq L\,|x - y|\,.$$

The Fourier series associated with a Lipschitz continuous converges uniformly and represents f, and since a $\mathcal{C}^1$ is, in particular, Lipschitz continuous, the uniform convergence of the Fourier series also holds for this class of functions.

Fourier series of $\cos(z\,x)$ and its consequences

The Fourier series of the function $f(x) = \cos(z\,x)$ (here z is a parameter) has, as we shall see, a great impact on our future work. Even if it is not 2π periodic, since it is an even function, its Fourier series converges uniformly to its periodic extension related to $[-\pi, \pi]$. To determine the coefficients we use again proposition 1.21, so that for $n \geq 1$

$$\begin{aligned}
a_n &= \frac{1}{\pi}\int_{-\pi}^{\pi} \cos(z\,x)\,\cos(n\,x)\,\mathrm{d}x = \frac{1}{\pi}\left[\frac{\sin(x(n-z))}{2(n-z)} + \frac{\sin(x(n+z))}{2(n+z)}\right]_{x=-\pi}^{x=\pi} \\
&= \frac{1}{\pi}\left(\frac{\sin(\pi(n-z))}{n-z} + \frac{\sin(\pi(n+z))}{n+z}\right) = \\
&= \frac{1}{\pi}\left(\frac{(-1)^{n+1}\sin(\pi\,z)}{n-z} + \frac{(-1)^{n}\sin(\pi\,z)}{n+z}\right) = \frac{(-1)^n}{\pi}\,\frac{2z}{z^2-n^2}\,\sin(\pi\,z).
\end{aligned}$$

While for $n = 0$ we have

$$a_0 = \frac{1}{\pi}\int_{-\pi}^{\pi} \cos(z\,x)\,\mathrm{d}x = \frac{2}{\pi\,z}\sin(\pi\,z)$$

Finally, because $f(x)$ is an even function, we have $b_n = 0$. Therefore, we can conclude that

Theorem. The Fourier series of $f(x) = \cos(z\,x)$ is given by

$$\cos(z\,x) = \frac{2\,z}{\pi}\sin(\pi\,z)\left(\frac{1}{2\,z^2} + \sum_{n=1}^{\infty}\frac{(-1)^n\cos(n\,x)}{z^2-n^2}\right). \tag{1.32}$$

Remark. In particular taking $x = 0$ we see that for $z \notin \mathbb{Z}$ we have

$$\frac{\pi}{\sin(\pi z)} = \frac{1}{z} - \sum_{n=1}^{\infty}(-1)^n \frac{2z}{z^2 - n^2} \tag{1.33}$$

Also taking $x = \pi$ in (1.32) we see that for $z \notin \mathbb{Z}$ we have

$$\cot(\pi z) = \frac{2}{\pi}\left(\frac{1}{2z} + \sum_{n=1}^{\infty} \frac{z}{z^2 - n^2}\right) \tag{1.34}$$

Identity (1.32) has many important consequences, among which, in addition to the reflection formula, there is the Gauss multiplication formula, Section 2.7, and Euler infinite product representation of the sine, which is the object of the next section and with the Bernoulli numbers and finally in the theory of zeta function, Theorem 4.2. Furthermore, using the Euler relation $\coth(ix) = i\cot(x)$ we also have the series expansion of the hyperbolic cotangent, which has a great impact on an important result, see Theorem 3.31 in chapter 3:

$$\coth(\pi z) = \frac{2}{\pi}\left(\frac{1}{2z} + \sum_{n=1}^{\infty} \frac{z}{z^2 + n^2}\right). \tag{1.35}$$

Consider $\sin(z\,x)$ in $[-\pi, \pi]$. Hence $a_n = 0$ and, assuming $z \notin \mathbb{Z}$

$$b_n = \frac{2}{\pi}\int_0^{\pi} \sin(zx)\sin(nx)\,\mathrm{d}x = (-1)^n \frac{2n\sin(z\,\pi)}{\pi\,(z^2 - n^2)}.$$

In conclusion, we have the series representation, valid for $z \notin \mathbb{Z}$:

$$\sin(z\,x) = \frac{2\sin(z\,\pi)}{\pi}\sum_{n=1}^{\infty}(-1)^n \frac{n}{z^2 - n^2}\sin(nx)\,.$$

Fourier series in arbitrary intervals

The choice of a different interval does not change the sufficient conditions for convergence stated in Theorems 1.22 and 1.23; the aspect that interests us here is related to the definition of the Fourier coefficients. Starting with $f : \mathbb{R} \to \mathbb{R}$ periodic of fundamental period $2L$, we can build, for $m \in \mathbb{N}$ the trigonometric series:

$$\frac{a_0}{2} + \sum_{n=1}^{\infty}\left[a_n \cos\left(\tfrac{n\,\pi\,x}{L}\right) + b_n \sin\left(\tfrac{n\,\pi\,x}{L}\right))\right], \tag{1.36}$$

where:

$$a_n = \frac{1}{L}\int_{-L}^{L} f(x)\cos\left(\tfrac{n\,\pi\,x}{L}\right)\,\mathrm{d}x, \quad b_n = \frac{1}{L}\int_{-L}^{L} f(x)\sin\left(\tfrac{n\,\pi\,x}{L}\right)\,\mathrm{d}x.$$

As usual, (a_n) starts from $n = 0$ while (b_n) starts from $n = 1$. Series (1.36) converges pointwise [uniformly] to f when the assumptions of Theorem 1.22 [1.23] are verified.

Complex form of Fourier series

Recalling Euler's formulas (1.16b) and (1.16c) we can represent a Fourier series in the complex form in the following way:

$$f(x)=\frac{a_0}{2}+\frac{1}{2}\sum_{n=1}^{\infty}\left((a_n-i\,b_n)\,e^{\frac{i\pi nx}{L}}+(a_n+i\,b_n)\,e^{-\frac{i\pi nx}{L}}\right).$$

The latter expression is represented in a more compact and elegant form as follows:

$$f(x)=\sum_{n=-\infty}^{\infty} c_n e^{\frac{i\,n\,x}{L}},\; c_0=\frac{a_0}{2};\; c_n=\frac{a_n-i\,b_n}{2},\; n>0;\; c_{-n}=\frac{a_n+i\,b_n}{2},\; n<0\,.$$

The bilateral sequence (c_n) terms can be calculated directly using the relation:

$$c_n=\frac{1}{2L}\int_{-L}^{L} f(x)\,e^{-\frac{i\,n\,x}{L}}\,\mathrm{d}x\,. \tag{1.37}$$

1.8 Euler infinite product for the Sine

Having in mind to demonstrate the formula for the representation of the sine by infinite product, we recall the main aspects of the theory of infinite products that have many points of contact with that of the series. We follow the exposition of [39], [77]. Let (b_n) be a complex sequence. The infinite product

$$\prod_{n=1}^{\infty} b_n = b_1\cdot b_2\cdot b_3\cdots \tag{1.38}$$

is said to be convergent is there exists $m\in\mathbb{N}$ such that $b_n\neq 0$ for any $n\geq m$ and the sequence of partial products:

$$\lim_{n\to\infty}\prod_{k=m}^{n} b_k=\lim_{n\to\infty}\,(b_m\cdot b_{m+1}\cdots b_n) \tag{1.39}$$

converges. Therefore, the value of the infinite product is

$$\prod_{n=1}^{\infty} b_n := b_1\cdot b_2\cdots b_{m-1}\cdot p\,.$$

The definition is independent of the chosen value of m for which the b_n terms are assumed nonzero for every $n\geq m$.

Infinite product (1.38) diverges if it does not converge, which means that either there exist infinite indices k for which $b_k=0$ or that the limit (1.39) diverges or converges to zero. In the latter case, we say that the infinite product (1.38) diverges to zero.

As in the case of infinite series, the first index from which to start an infinite product can be any integer, the meaning of

$$\prod_{n=k}^{\infty} b_n$$

is straightforward. We can state a necessary convergence condition in analogy with the theory of the infinite series.

Theorem. If an infinite product converges, then its factors tend to 1. Also, an infinite product has a value of 0 if and only if it has a null factor.

It is convenient to write the sequence of factors in the form $b_n = 1 + a_n$ so that the infinite product (1.38) can be written as:

$$\prod_{n=1}^{\infty} (1 + a_n) \,. \tag{1.40}$$

Hence the convergence of the infinite product (1.40) implies that $a_n \to 0$.

When the sequence (a_n) appearing in (1.40) is real and nonnegative we have the following:

Theorem. An infinite product represented in the form (1.40) with non-negative terms a_n converges if and only if the series

$$\sum_{n=1}^{\infty} a_n$$

converges.

In the case of real sequences, but not necessarily non-negative, the following result applies, for which we refer to [77, pp. 357–358].

Theorem. Given (a_n) the infinite product (1.38) converges if and only if $a_n \to 0$ and the series

$$\sum_{n=m+1}^{\infty} \ln(1 + a_n)$$

converges. Moreover, if s denotes the sum of the infinite series, then:

$$\prod_{n=1}^{\infty}(1 + a_n) = (1 + a_1) \cdots (1 + a_m) e^s \,.$$

Sine infinite product

The representation of the sine using an infinite product was established by Euler in 1735, in the process that led him to his first solution of the Basel problem, even if rigorous proof was provided by Weierstrass one century later. In the years following this, several demonstrations of this remarkable formula have been produced, which will prove extremely useful in our study of the Gamma function in chapter 2. We will provide proof based on (1.34), which was derived from the Fourier expansion of $\cos(z\,x)$, equation (1.32).

Theorem. The following formula, which represents the sine function in terms of an infinite product, holds:

$$\sin x = x \prod_{n=1}^{\infty} \left(1 - \frac{x^2}{n^2 \pi^2}\right). \tag{1.41}$$

Proof. In order to demonstrate (1.41) we begin observing that by looking at (1.34) we see that convergence is uniform in all compact intervals $[-\alpha, \alpha]$ with $0 < \alpha < 1$, thus it is permissible to integrate (1.32) in an interval of the form $[\varepsilon, x]$ being $0 < \varepsilon < x < 1$ obtaining:

$$\frac{1}{\pi} \ln \left(\frac{\sin(\pi\, x)}{\sin(\pi\, \varepsilon)}\right) = \frac{1}{\pi} \ln \left(\frac{x}{\varepsilon}\right) + \frac{1}{\pi} \sum_{n=1}^{\infty} \ln \left(\frac{x^2 - n^2}{\varepsilon^2 - n^2}\right).$$

Now, after simplifying the factor $1/\pi$, if we take the first addend from the right-hand side to the left-hand side, we obtain

$$\ln \left(\frac{\sin(\pi\, x)}{x} \, \frac{\varepsilon}{\sin(\pi\, \varepsilon)}\right) = \sum_{n=1}^{\infty} \ln \left(\frac{x^2 - n^2}{\varepsilon^2 - n^2}\right).$$

We can take the limit for $\varepsilon \to 0^+$ to infer that, for any $0 < x < 1$:

$$\ln \left(\frac{\sin(\pi\, x)}{\pi\, x}\right) = \sum_{n=1}^{\infty} \ln \left(\frac{n^2 - x^2}{n^2}\right).$$

(1.41) follows exponentiating the last equality and replacing $\pi\, x$ with x. □

Remark. From (1.41) follows a similar relationship for cosine:

$$\cos x = \prod_{n=1}^{\infty} \left(1 - \frac{4x^2}{\pi^2 (2n-1)^2}\right). \tag{1.41b}$$

1.9 Fourier transform

We conclude our brief review of the notions that we will need in our study of Special Functions by providing the minimal but indispensable notions that will be used in the following on the Fourier and Laplace transforms, which will be the object of the next section.

Definition. Let $f \in \mathcal{L}^1(\mathbb{R})$ be a function of a real variable. The Fourier transform of f is the complex-valued function:

$$\mathcal{F}f(s) := \int_{-\infty}^{+\infty} e^{-2\pi i\, st} f(t)\, \mathrm{d}t. \tag{1.42}$$

Remark. The Fourier transform is a bounded linear operator, from the linear space $\mathcal{L}^1(\mathbb{R})$ to the linear space of continuous functions $\mathcal{C}(\mathbb{R})$. Moreover,

$$\lim_{|s|\to\infty} \mathcal{F}f(s) = 0.$$

Linearity is a straightforward consequence of the properties of the integral. The continuity of $\mathcal{F}f(s)$ follows from the following argument: fixed $s \in \mathbb{R}$ take $h \in \mathbb{R}$ and consider, as usual

$$\begin{aligned} |\mathcal{F}f(s+h) - \mathcal{F}f(s)| &= \left| \int_{-\infty}^{\infty} \left(1 - e^{2i\pi ht}\right) e^{-2i\pi t(h+s)} f(t) \right| \mathrm{d}t \\ &\le \int_{-\infty}^{\infty} \left|1 - e^{2i\pi ht}\right| |f(t)|\, \mathrm{d}t\,. \end{aligned}$$

The right-hand side goes to zero for $h \to 0$, proving the continuity of $\mathcal{F}f(s)$. Finally, it is clear that

$$\sup_{s\in\mathbb{R}} |\mathcal{F}f(s)| \le \int_{-\infty}^{\infty} |f(t)|\, \mathrm{d}t = \|f\|_1\,. \tag{1.43}$$

This proves the boundedness of the Fourier integral operator. Notice that $\|f\|_1$ is the $\mathcal{L}^1$ norm. Finally, having observed that using a translation, we have the following:

$$\mathcal{F}f(s) = \int_{-\infty}^{\infty} e^{-2\pi i\, s(t+h)} f(t+h)\, \mathrm{d}t = e^{-2\pi i\, sh} \int_{-\infty}^{\infty} e^{-2\pi i\, st} f(t+h)\, \mathrm{d}t\,.$$

Now we can choose h so that $e^{-2\pi i\, sh} = -1$ in this way we have

$$\mathcal{F}f(s) = -\int_{-\infty}^{\infty} e^{-2\pi i\, st} f(t + \frac{1}{2s})\, \mathrm{d}t\,.$$

Thus, adding side by side the latter with (1.42), we arrive at

$$2\mathcal{F}f(s) = \int_{-\infty}^{\infty} e^{-2\pi i\, st} \left(f(t) - f(t + \frac{1}{2s}) \right) \mathrm{d}t\,.$$

The right-hand side of the last equality goes to zero due to the properties of mean continuity of $\mathcal{L}^p$ spaces.

Remark. The Fourier transform of a summable function is not necessarily summable. In fact

$$f(t) = \operatorname{sgn} t\, e^{-|t|} \implies \mathcal{F}f(s) = -\frac{4\pi\, s}{1 + 4\pi^2 s^2}\, i\,.$$

The next remark is of some interest.

Remark. If $f,\, g \in \mathcal{L}^1(\mathbb{R})$ then

$$\int_{-\infty}^{\infty} \mathcal{F}f(s)\, g(s)\, \mathrm{d}s = \int_{-\infty}^{\infty} f(t)\, \mathcal{F}g(t)\, \mathrm{d}t. \tag{1.44}$$

Notice that both integrals in (1.44) converge because Fourier transforms is bounded; we can, therefore, exchange the order of integration on the left-hand side of (1.44) thanks to the exchange of the order of integration as follows:

$$\begin{aligned}
\int_{-\infty}^{\infty} \mathcal{F}f(s)\, g(s)\, \mathrm{d}s &= \int_{-\infty}^{\infty} g(s) \left(\int_{-\infty}^{+\infty} e^{-2\pi\, i\, st}\, f(t)\, \mathrm{d}t \right) \mathrm{d}s \\
&= \int_{-\infty}^{\infty} f(t) \left(\int_{-\infty}^{+\infty} e^{-2\pi\, i\, st}\, g(s)\, \mathrm{d}s \right) \mathrm{d}t \\
&= \int_{-\infty}^{\infty} f(t)\, \mathcal{F}g(t)\, \mathrm{d}t\,.
\end{aligned}$$

We analyse the behaviour of odd and even functions concerning the Fourier transform.

Remark. If f is an even function i.e. $f(t) = f(-t)$, then

$$\mathcal{F}f(s) = \int_{-\infty}^{+\infty} \cos(2\pi st)\, f(t)\, \mathrm{d}t = 2 \int_{0}^{+\infty} \cos(2\pi st)\, f(t)\, \mathrm{d}t\,.$$

If f is an odd function, i.e. $f(t) = -f(-t)$, then

$$\mathcal{F}f(s) = -i \int_{-\infty}^{+\infty} \sin(2\pi st)\, f(t)\, \mathrm{d}t = -2i \int_{0}^{+\infty} \sin(2\pi st)\, f(t)\, \mathrm{d}t\,.$$

Fourier transform of the Gaussian

The density of the Gaussian normal probability distribution is a fixed point for the Fourier transform; this fact has many important applications. Among these, some are closely interested in our work on special functions.

If we consider the function $f(t) = e^{-\pi\, t^2}$ we see that $\mathcal{F}f(s) = e^{-\pi\, s^2}$. The first step to be taken is to calculate the integral explicitly

$$p = \int_{-\infty}^{\infty} f(t)\, \mathrm{d}t = \int_{-\infty}^{\infty} e^{-\pi\, t^2}\, \mathrm{d}t.$$

Although the calculation of this integral will be presented in chapter 2, equation (2.31b), by exploiting the properties of Eulerian integrals of the first and second kind, we perform its calculation here using a very popular method for completeness of exposition. Consider the double integral:

$$q = \iint_{\mathbb{R}^2} e^{-\pi(t^2+s^2)}\mathrm{d}t\, \mathrm{d}s.$$

A straightforward application of the Fubini reduction formula yields recalling the fundamental property of the exponential:

$$q = \int_{-\infty}^{\infty} e^{-\pi\, t^2}\, \mathrm{d}t \int_{-\infty}^{\infty} e^{-\pi\, s^2}\, \mathrm{d}s = p^2 .$$

On the other hand, q can be calculated using the polar coordinates $t = \rho \cos\vartheta$, $s = \rho \sin\vartheta$:

$$q = \iint_{[0,2\pi)\times[0,\infty)} \rho\, e^{-\pi\rho^2}\, \mathrm{d}\rho\, \mathrm{d}\vartheta = 2\pi \int_0^{\infty} \rho\, e^{-\pi\rho^2}\, \mathrm{d}\rho = 1.$$

Ultimately, we have confirmed that the function $f(t)$ is indeed a probability density, being

$$p = \int_{-\infty}^{\infty} f(t)\, \mathrm{d}t = \sqrt{q} = 1.$$

Let us determine the Fourier transform of $f(t)$. The transform can be evaluated by completing the square in the exponent. We illustrate this procedure in full detail.

$$\mathcal{F}f(s) = \int_{-\infty}^{+\infty} e^{-2\pi i s t}\, e^{-\pi t^2}\, \mathrm{d}t = \int_{-\infty}^{+\infty} e^{-\pi\,(2\,i\,s\,t + t^2)}\, \mathrm{d}t.$$

Now, rewrite the exponent as

$$-\pi\left(2ist + t^2\right) = -\pi(-s^2 + 2ist + t^2 + s^2) = -\pi s^2 - \pi(t + i\,s)^2 .$$

Thus

$$\mathcal{F}f(s) = e^{-\pi\, s^2} \int_{-\infty}^{+\infty} e^{-\pi\,(t+i\,s)^2}\, \mathrm{d}t\, .$$

Employ the change of variable $(t + i\,s) = \tau$ so that

$$\mathcal{F}f(s) = e^{-\pi s^2} \int_{-\infty}^{+\infty} e^{-\pi\tau^2}\, \mathrm{d}\tau.$$

Hence

$$\mathcal{F}f(s) = e^{-\pi s^2}\, .$$

We are proving our statement.

Later in this section, we will need a property of the transformation known as the stretch property.

Proposition 1.24. If $a \in \mathbb{R}$, $a \neq 0$ and $f : \mathbb{R} \to \mathbb{R}$, define ${}_a f(t) = f(a\,t)$. Then

$$\mathcal{F}_a f(s) = \frac{1}{|a|}\, \mathcal{F}f\left(\frac{s}{a}\right) .$$

Proof. First, assume that $a > 0$. Then

$$\mathcal{F}_a f(s) = \int_{-\infty}^{+\infty} f(a\,t)\, e^{-2\pi ist}\,\mathrm{d}t = \frac{1}{a}\int_{-\infty}^{+\infty} f(u)\, e^{-2\pi is(u/a)}\,\mathrm{d}u$$
$$= \frac{1}{a}\int_{-\infty}^{+\infty} f(u)\, e^{-2\pi iu(s/a)}\,\mathrm{d}u = \frac{1}{a}\mathcal{F}f\left(\frac{s}{a}\right).$$

When $a < 0$, the limits of integration are reversed when we insert the substitution $u = at$, thus, the resulting transform is

$$\mathcal{F}_a f(s) = -\frac{1}{a}\mathcal{F}f\left(\frac{s}{a}\right).$$

Since $-a$ is positive when a is negative, We can combine the two cases to end our argument.

□

Convolution

Convolution is an operation that combines two functions, f, g producing a third function as shown in the following formula (1.45).

Definition. The convolution of two functions $g(t)$ and $f(t)$, both defined on the entire real line, is defined as:

$$(g \star f)(t) = \int_{-\infty}^{+\infty} g(t-x)\; f(x)\;\mathrm{d}x\,. \tag{1.45}$$

Convolution behaves nicely concerning the Fourier transform, as shown by the following result, known as the Convolution Theorem for the Fourier transform.

Theorem. If $g(t)$ and $f(t)$ are both summable, then

$$\mathcal{F}(g \star f)(s) = \mathcal{F}g(s)\,\mathcal{F}f(s).$$

Proof. The proof is straightforward and uses the exchange of the order of integration:

$$\mathcal{F}(g \star f)(s) = \int_{-\infty}^{\infty} (g \star f)(t)\; e^{-2\pi\, ist}\,\mathrm{d}t$$
$$= \int_{-\infty}^{\infty}\left(\int_{-\infty}^{\infty} g(t-x)\, f(x)\, e^{-2\pi ist}\,\mathrm{d}x\right)\,\mathrm{d}t$$
$$= \int_{-\infty}^{\infty}\left(\int_{-\infty}^{\infty} g(t-x)\, f(x)\, e^{-2\pi ist}\,\mathrm{d}t\right)\,\mathrm{d}x$$
$$= \int_{-\infty}^{\infty} f(x)\left(\int_{-\infty}^{\infty} g(t-x)\, e^{-2\pi ist}\,\mathrm{d}t\right)\,\mathrm{d}x.$$

With the change of variable $t - x = u$, so that $\mathrm{d}t = \mathrm{d}u$, we obtain

$$\begin{aligned}\mathcal{F}(g \star f)(s) &= \int_{-\infty}^{\infty} f(x) \left(\int_{-\infty}^{\infty} g(u)\, e^{-2\pi i s\,(u+x)}\, \mathrm{d}u \right) \mathrm{d}x \\ &= \int_{-\infty}^{\infty} f(x)\, e^{-2\pi i s x} \left(\int_{-\infty}^{\infty} g(u)\, e^{-2\pi i s u}\, \mathrm{d}u \right) \mathrm{d}x \\ &= \mathcal{F}g(s) \int_{-\infty}^{\infty} f(x)\, e^{-2\,\pi i s x}\, \mathrm{d}x = \mathcal{F}g(s)\, \mathcal{F}f(s).\end{aligned}$$

□

Schwartz space

Given what is, for our purposes, the main application of the Fourier transform, we need to operate on a class of smooth and rapidly decreasing functions.

Definition (Schwartz space)**.** The Schwartz space $\mathcal{S}(\mathbb{R})$ is the vector space of the functions $f : \mathbb{R} \to \mathbb{C}$ defined as

$$\mathcal{S}(\mathbb{R}) = \left\{ f \in \mathcal{C}^{\infty}(\mathbb{R}) : \text{ for any } n,\, m \in \mathbb{N},\ \sup_{x \in \mathbb{R}} \left| x^n f^{(m)}(x) \right| < +\infty \right\}. \tag{1.46}$$

It is also called the space of rapidly decreasing functions, since $\mathcal{S}(\mathbb{R})$ is the function space of all functions whose derivatives are rapidly decreasing to zero.

Remark. The space $\mathcal{S}(\mathbb{R})$ can be characterised equivalently by assuming that for each $n,\, m \in \mathbb{N}$ the following equality holds:

$$\lim_{|x| \to \infty} \left| x^n f^{(m)}(x) \right| = 0\,. \tag{1.47}$$

Indeed if we assume (1.46), then there exists $c_{n,\,m} \in \mathbb{R}^+$ such that

$$\left| x^{n+1} f^{(m)}(x) \right| \le c_{n,\,m}$$

so we infer (1.47), in fact

$$\left| x^n f^{(m)}(x) \right| \le \frac{c_{n,\,m}}{|x|} \to 0 \text{ as } |x| \to +\infty.$$

The main characteristic of the space $\mathcal{S}(\mathbb{R})$ is that if $f \in \mathcal{S}(\mathbb{R})$ then $\mathcal{F} \in \mathcal{S}(\mathbb{R})$. To prove this fact, we must first determine the derivatives of the transform.

Theorem. If $f \in \mathcal{S}^1(\mathbb{R})$, then

$$\mathcal{F}f'(s) = 2\,\pi i s\, \mathcal{F}f(s) \tag{1.48a}$$
$$\mathcal{F}'(s) = \mathcal{F}g(s) \tag{1.48b}$$

where $g(t) = -2\pi\, i\, t\, f(t)$.

Proof. To obtain (1.48a), it suffices to evaluate the Fourier transform of $f'(t)$:

$$\mathcal{F}f'(s) = \int_{-\infty}^{+\infty} e^{-2\pi ist} f'(t)\,\mathrm{d}t$$

and integrate by parts. Formula (1.48b) follows differentiating the Fourier integral, which is correct since $f \in \mathcal{S}(\mathbb{R})$, indeed

$$\frac{\mathrm{d}}{\mathrm{d}s}\mathcal{F}(s) = \frac{\mathrm{d}}{\mathrm{d}s}\int_{-\infty}^{+\infty} e^{-2\pi ist} f(t)\,\mathrm{d}t = \int_{-\infty}^{+\infty} -2\pi\, i\, t\, e^{-2\pi ist} f(t)\,\mathrm{d}t.$$

□

Note that differentiation is transformed into multiplication: This represents another remarkable feature of the Fourier transform, providing one important reason for its usefulness. Formulas for higher derivatives also hold, as well as the relevant result follows by mathematical induction:

$$\mathcal{F}f^{(n)}(s) = (2\,\pi\, i\, s)^n\ \mathcal{F}f(s) \tag{1.49a}$$

$$\mathcal{F}^{(n)}(s) = \mathcal{F}g_n(s) \tag{1.49b}$$

where $g_n(t) = (-2\pi\, i\, t)^n f(t)$.

The fundamental property of the Schwartz space related to Fourier transform follows from (1.48) and (1.49).

Theorem. If $f \in \mathcal{S}(\mathbb{R})$ then $\mathcal{F}f \in \mathcal{S}(\mathbb{R})$

Proof. Given $f \in \mathcal{S}(\mathbb{R})$ let n, $m \in \mathbb{N}$ and define $g_{n,m}(x)$ as

$$g_{n,m}(x) = \left((-2\pi\, i\, x)^m f(x)\right)^{(n)}$$

Combining the formulas (1.49), we infer

$$\mathcal{F}g_{n,m}(s) = (2\pi\, i\, s)^n \mathcal{F}^{(m)}(s) \tag{1.50}$$

To prove our statement, we have to show that there exists $c > 0$ such that

$$\left|s^n \mathcal{F}^{(m)}(s)\right| \leq c \tag{1.51}$$

Using (1.50) we see that

$$|s^n|\ \left|\mathcal{F}^{(m)}(s)\right| = (2\pi)^{m-n}\, |\mathcal{F}g_{n,m}(s)| \leq (2\pi)^{m-n}\, \|g_{n,m}\|_1$$

where we used (1.43) and the fact that $f \in \mathcal{S}(\mathbb{R})$ implies $g_{n,m} \in \mathcal{L}^1$ hence (1.51)is proved and thus the proof is completed. □

Poisson summation formula

Operating in Schwartz space, it is possible to demonstrate the Poisson's summation formula used in studying the zeta function, which we will discuss in chapter 3.

Theorem 1.25. If $f \in \mathcal{S}(\mathbb{R})$ then the function:

$$F(x) = \sum_{n\in\mathbb{Z}} f(x+n),$$

is well-defined and periodic with period 1. Moreover, Poissons summation formula (1.52) holds:

$$\sum_{n\in\mathbb{Z}} f(n) = \sum_{n\in\mathbb{Z}} \mathcal{F}f(n). \tag{1.52}$$

Proof. The assumption $f \in \mathcal{S}(\mathbb{R})$ implies that series-defining $F(x)$ is convergent as well as both sides of (1.52). The periodicity of $F(x)$ follows from the fact that, being

$$F(x+1) = \sum_{n\in\mathbb{Z}} f(x+1+n)$$

changing index putting $m = n+1$ observed that when n varies in $\mathbb{Z}$ also m spans $\mathbb{Z}$ we see that

$$F(x+1) = \sum_{m\in\mathbb{Z}} f(x+m) = F(x).$$

At this point, given the periodicity of $F(x)$, it is natural to determine its Fourier series; we have, from (1.37), recalling that in this case we have $L = 1/2$:

$$F(x) = \sum_{n\in\mathbb{Z}} c_n e^{2\pi\, i\, n x}, \tag{1.53}$$

where

$$c_n = \int_0^1 F(x)\, e^{-2\pi\, i\, n x} \mathrm{d}x.$$

The calculation of the coefficients c_n naturally leads to the Fourier transform; in fact, plugging the expression defining F in the integral we find, notice that having $f \in \mathcal{S}(\mathbb{R})$ allows the exchange between integral and infinite series:

$$\begin{aligned} c_n &= \int_0^1 \sum_{m\in\mathbb{Z}} f(x+m)\, e^{-2\pi\, i\, n x} \mathrm{d}x = \sum_{m\in\mathbb{Z}} \int_0^1 f(x+m)\, e^{-2\pi\, i\, n x} \mathrm{d}x \\ &= \sum_{m\in\mathbb{Z}} \int_m^{m+1} f(y)\, e^{-2\pi\, i\, n(y-m)} \mathrm{d}y = \sum_{m\in\mathbb{Z}} e^{2i\pi\, mn} \int_m^{m+1} f(y)\, e^{-2i\pi\, ny} \mathrm{d}y \\ &= \sum_{m\in\mathbb{Z}} \int_m^{m+1} f(y)\, e^{-2i\pi\, ny} \mathrm{d}y. \end{aligned}$$

Note that in the third step, we used the change of variable $y = x + m$ in the integral, in the fourth the property of the exponential, and the last, the fact that being $n, m \in \mathbb{Z}$ then $e^{2i\pi\, mn} = 1$. Finally, using the additivity of the integral, we arrive at

$$c_n = \int_{-\infty}^{\infty} f(y)\, e^{-2i\pi\, ny} \mathrm{d}y = \mathcal{F}f(n).$$

Therefore, by inserting the Fourier coefficients in (1.53) we arrive, at recalling the definition of $F(x)$, at

$$\sum_{n\in\mathbb{Z}} f(x+n) = \sum_{n\in\mathbb{Z}} \mathcal{F}f(n)\, e^{2\pi\, i\, nx} \tag{1.53b}$$

Equation (1.52) follows from (1.53b) taking $x = 0$. □

1.10 Jacobi theta function

The consequence of the Poisson summation formula (1.52) that interests our work on special functions concerns precisely a transcendent function that was introduced by Jacobi in 1829:

Definition. The Jacobi theta function is defined as

$$\theta(x) = \sum_{n\in\mathbb{Z}} e^{-\pi\, x\, n^2}. \tag{1.54}$$

Suppose x is a real positive parameter. In that case, $f(n) = e^{-\pi x n^2}$ is an element of $\mathcal{S}(\mathbb{R})$: this allows us to use Theorem 1.25 and formulate the following fundamental statement, relating to the functional property of θ.

Theorem. If $x > 0$, the Jacobi theta function satisfies the functional equation

$$\theta(x) = \frac{1}{\sqrt{x}}\, \theta\left(\tfrac{1}{x}\right) \tag{1.55}$$

Proof. We already observed that $f \in \mathcal{S}(\mathbb{R})$, moreover, the Fourier transform $\mathcal{F}f(\nu)$ of $f(n)$ is, in force of proposition 1.24

$$\mathcal{F}f(\nu) = \frac{e^{-\frac{\pi\nu^2}{x}}}{\sqrt{x}}.$$

Hence, we can invoke Theorem 1.25 and infer

$$\theta(x) = \sum_{n\in\mathbb{Z}} e^{-\pi\, x\, n^2} = \sum_{\nu\in\mathbb{Z}} \frac{e^{-\frac{\pi\nu^2}{x}}}{\sqrt{x}} = \frac{1}{\sqrt{x}}\, \theta\left(\tfrac{1}{x}\right).$$

Remark. The formula (1.54) can be rewritten as

$$\theta(x) = 1 + 2\sum_{n=1}^{\infty} e^{-\pi x n^2} = 1 + 2\,\Psi(x) \tag{1.54b}$$

where, of course:

$$\Psi(x) = \sum_{n=1}^{\infty} e^{-\pi x n^2}. \tag{1.56}$$

In terms of Ψ functional equation (1.55) becomes

$$1 + 2\Psi(x) = \frac{1}{\sqrt{x}}\left(2\Psi\left(\frac{1}{x}\right) + 2\right)$$

so that, solving for $\Psi(x)$ we obtain

$$\Psi(x) = \frac{1}{\sqrt{x}}\Psi\left(\frac{1}{x}\right) + \frac{1}{2\sqrt{x}} - \frac{1}{2}. \tag{1.57}$$

□

1.11 Laplace transform

Definition. $f : [0, +\infty) \to \mathbb{R}$ is said Laplace transformable if the integral

$$\int_0^{+\infty} e^{-sx} f(x)\,\mathrm{d}x$$

converges for some $s \in \mathbb{R}$. In this case, the Laplace transform of f is given by

$$\mathcal{L}[f](s) := \int_0^{+\infty} e^{-sx} f(x)\,\mathrm{d}x. \tag{L}$$

We will also write $F(s)$ to denote the Laplace transform.

Remark. Integral (L) can be thought of as a generalised Riemann integral, in the sense that we assume $t \mapsto f(t)e^{-st}$ Riemann integrable on each interval $[0, b]$ for any $b > 0$ and we define

$$\int_0^{+\infty} e^{-sx} f(x)\,\mathrm{d}x := \lim_{b\to+\infty} \int_0^{b} e^{-sx} f(x)\,\mathrm{d}x,$$

when the limit on the right-hand side is finite.

The fact that we have yet to use the Lebesgue integral should not lead to fear of a loss of generality since the class of functions on which we will focus is either continuous functions or with a finite number of jump discontinuities, for which the two notions of integral coincide. If for $s = 0$ the Laplace transform is defined, i.e. f is integrable in $[0, +\infty)$, its value is

$$\mathcal{L}[f](0) = F(0) = \int_0^{+\infty} f(x)\,\mathrm{d}x.$$

The transformed function $F(s)$ domain is formed by all the values of s for which the integral (L) converges; we denote this set with the usual notation $\operatorname{dom} F$. The following theorem illustrates the nature of $\operatorname{dom} F$.

Theorem 1.26. If f is Laplace transformable and $s_0 \in \operatorname{dom} F$ then $s \in \operatorname{dom} F$ for any $s > s_0$.

Proof. Recall the Cauchy convergence criterion for generalised integrals: if $g : [0, +\infty) \to \mathbb{R}$ is integrable, then for any $\varepsilon > 0$ there exists $\xi > 0$ such that for any $x_1,\ x_2 \in [0, +\infty)$ such that $x_1,\ x_2 > \xi$ then

$$\left| \int_{x_1}^{x_2} g(x)\,\mathrm{d}x \right| < \varepsilon.$$

Thus if $s_0 \in \operatorname{dom} F$ choose $s > s_0$ so that

$$\begin{aligned}
\left| \int_{x_1}^{x_2} f(x) e^{-sx}\,\mathrm{d}x \right| &= \left| \int_{x_1}^{x_2} f(x) e^{-(s-s_0)x} e^{-s_0\,x}\,\mathrm{d}x \right| \\
&\le \int_{x_1}^{x_2} \left| f(x) e^{-(s-s_0)x} e^{-s_0\,x} \right| \mathrm{d}x \\
&= \int_{x_1}^{x_2} e^{-(s-s_0)x} \left| f(x) e^{-s_0\,x} \right| \mathrm{d}x \le \int_{x_1}^{x_2} \left| f(x) e^{-s_0 x} \right| \mathrm{d}x.
\end{aligned}$$

Since the last integral obtained is by hypothesis convergent, the thesis follows from the aforementioned Cauchy convergence criterion. □

As a consequence of the Theorem 1.26 we see that if there exists s_0 for which the Laplace integral converges $\operatorname{dom} F$ is an interval of the type $[\sigma_f, +\infty)$ or $(\sigma_f, +\infty)$ where $\sigma_f := \inf\{s;\ s \in \operatorname{dom} F\}$. In this case σ_f is said *convergence abscissa*. In the degenerate case, where the integral (L) does not converge for any s_0, then $\operatorname{dom} F = \emptyset$.

Example 1.27. We compute the transform when $f(x) = 1$ for any $x \ge 0$ even if it is one of the simplest examples; it illustrates how the problem of determining the abscissa of convergence arises in a completely natural way: by definition, we have

$$F(s) = \int_0^{+\infty} e^{-sx}\mathrm{d}x\,.$$

Here is we have $s > 0 := s_f$, si that, the transform is given by

$$F(s) = \lim_{b \to +\infty} \left[-\frac{e^{-sx}}{s} \right]_{x=0}^{x=b} = \frac{1}{s}.$$

The convergence interval is open being $\operatorname{dom} F = (0, +\infty)$. The interval of converge remains the same for $f(x) = x$, while the transform is given by

$$F(s) = \lim_{b \to +\infty} \left[-\frac{e^{-sx}(sx+1)}{s^2} \right]_{x=0}^{x=b} = \frac{1}{s^2}.$$

The convergence interval can contain its lower bound; for instance, consider:

$$f(x) = \begin{cases} 0 & \text{if } 0 \leq x \leq 2 \\ \dfrac{e^{3x}}{x\sqrt{x}} & \text{if } x > 2 \end{cases}$$

Hence, its transformation leads to the integral.

$$F(s) = \int_2^{+\infty} \frac{e^{(3-s)x}}{x\sqrt{x}}\, \mathrm{d}x,$$

which obviously converges for $s \in \operatorname{dom} F = [3, +\infty)$. However, it is impossible to express the transformation in terms of elementary functions: in this case, the error functions are needed. We conclude these preliminary considerations by noting that some functions, such as $f(x) = e^{x^2}$, are not transformable. In this case, no real values of s lead to the transform's convergence.

Having directly ascertained that only some functions are transformable suggests identifying sufficient conditions to ensure their transformability. At most, the behaviour at infinity must grow exponentially to ensure the considered function is effectively transformable.

Definition 1.28. $f : [0, +\infty) \to \mathbb{R}$ is said of exponential order if there exist $M > 0$ and $\tau \in \mathbb{R}$ such that for any $x \in [0, +\infty)$ $|f(x)| \leq M\, e^{\tau\, x}$. Moreover,

$$\tau_f := \inf \left\{ \tau \in \mathbb{R} : \exists\, M \geq 0 \, \text{such that } \; |f(x)| \leq M\, e^{\tau\, x} \text{ for any } x \in [0, +\infty) \right\}.$$

We now recall the notion of a piecewise continuous function.

Definition 1.29. $f : [0, +\infty) \to \mathbb{R}$ is said piecewise continuous if it has, at most, a finite number of jump discontinuity and is continuous elsewhere.

Therefore, the class of transformable functions on which we will focus our attention is those that satisfy both definitions 1.28 and 1.29, constituted by the functions of exponential order and piecewise continuous. It is not difficult to verify that this set constitutes a vector space and that polynomials, trigonometric and hyperbolic functions are indeed Laplace transformable. We will indicate this set with the $\mathcal{T}$.

When a function is transformable, its transform vanishes for $s \to +\infty$, as shown in the following statement.

Theorem. If $f \in \mathcal{T}$ such that f is also integrable on $[0,+\infty)$ then $\lim_{s\to+\infty} \mathscr{L}[f](s) = 0$.

Proof. Recalling that f is of exponential order there exist constants M and τ such that, if $s > \tau$

$$|\mathscr{L}[f](s)| = \left|\int_0^{+\infty} f(x)e^{-sx}\mathrm{d}x\right| \le \int_0^{+\infty} |f(x)|\, e^{-sx}\mathrm{d}x$$
$$\le \int_0^{+\infty} Me^{\tau\, x}\, e^{-sx}\mathrm{d}x = \frac{M}{s-\tau}$$

so, the thesis follows, taking the limit as $s \to +\infty$. □

It is common to represent the association between a function and its transform with the symbol $f(x) \circ\!\!-\!\!\bullet F(s)$

Example.

$$f(x) = 1 \quad \circ\!\!-\!\!\bullet \quad F(s) = \frac{1}{s}$$
$$f(x) = x \quad \circ\!\!-\!\!\bullet \quad F(s) = \frac{1}{s^2}$$
$$f(x) = x^n \quad \circ\!\!-\!\!\bullet \quad F(s) = \frac{n!}{s^{n+1}}$$
$$f(x) = e^{ax} \quad \circ\!\!-\!\!\bullet \quad F(s) = \frac{1}{s-a}$$
$$f(x) = \sin(ax) \quad \circ\!\!-\!\!\bullet \quad F(s) = \frac{a}{s^2+a^2}$$
$$f(x) = \cos(ax) \quad \circ\!\!-\!\!\bullet \quad F(s) = \frac{s}{s^2+a^2}$$
$$f(x) = \sinh(ax) \quad \circ\!\!-\!\!\bullet \quad F(s) = \frac{a}{s^2-a^2}$$
$$f(x) = \cosh(ax) \quad \circ\!\!-\!\!\bullet \quad F(s) = \frac{s}{s^2-a^2}$$

We have already shown the first two equalities; the third follows by iterating the integration by parts. Regarding the fourth, we have

$$\mathscr{L}[e^{ax}](s) = \int_0^{\infty} e^{ax}e^{-sx}\mathrm{d}x = \int_0^{\infty} e^{(a-s)x}\mathrm{d}x = \frac{1}{s-a},$$

where of course we must take $s > a$ that is $\sigma_{e^{ax}} = a$. Moreover,

$$\mathscr{L}[\sin(ax)](s) = \lim_{b\to+\infty}\int_0^{b} \sin(ax)\, e^{-sx}\mathrm{d}x$$
$$= \lim_{b\to+\infty}\left[-\frac{e^{-sx}(s\sin(ax)+a\cos(ax))}{a^2+s^2}\right]_{x=0}^{x=b} = \frac{a}{s^2+a^2}$$
$$\mathscr{L}[\cos(ax)](s) = \lim_{b\to+\infty}\int_0^{b} \cos(ax)\, e^{-sx}\mathrm{d}x$$
$$= \lim_{b\to+\infty}\left[\frac{e^{-sx}(a\sin(ax)-s\cos(ax))}{a^2+s^2}\right]_{x=0}^{x=b} = \frac{s}{s^2+a^2}$$

We must assume $s > 0$ in both cases to ensure integral convergence. To deal with the case of hyperbolic functions, we first observe that the transform is a linear operator, that is

$$\mathscr{L}\left[c_1 f_1 + c_2 f_2\right](s) = c_1 \mathscr{L}\left[f_1\right](s) + c_2 \mathscr{L}\left[f_2\right](s)$$

Thus, the two last statements follow using linearity and the definitions of hyperbolic sine and cosine:

$$\sinh(ax) = \frac{e^{ax} - e^{-ax}}{2}, \quad \cosh(ax) = \frac{e^{ax} + e^{-ax}}{2}$$

In both cases, we assume $s > |a|$.

The Laplace transform has interesting properties when applied to functions whose argument is translated. In this regard, it makes sense to introduce the so-called Heaviside step function.

$$h(x) = \begin{cases} 1 & \text{if } x \geq 0 \\ 0 & \text{if } x < 0 \end{cases}$$

The value assigned to the function in zero is immaterial. Some authors give the value $1/2$ in the origin: in any case, as is well known, the modification of the value of a function in a single point does not alter the value of its integral.

We present the first translation property here, which is sometimes said to be the first delay theorem when the independent variable x represents time.

Theorem. If $f \in \mathcal{T}$ such that f is also integrable on $[0, +\infty)$ and if $a > 0$ then

$$\mathscr{L}\left[f(x-a)h(x-a)\right](s) = e^{-as}\mathscr{L}\left[f(x)\right](s)$$

Proof. First, we follow the Laplace transform definition, and then we use the change of variable $x = y + a$ in such a way

$$\mathscr{L}\left[f(x-a)h(x-a)\right](s) = \int_a^{+\infty} f(x-a)e^{-sx}\,\mathrm{d}x = \int_0^{+\infty} f(y)e^{-s(y+a)}\,\mathrm{d}y$$
$$= e^{-sa}\int_0^{+\infty} f(y)e^{-sy}\,\mathrm{d}y = e^{-sa}\mathscr{L}\left[f(x)\right](s)\,.$$

□

For instance if $f(x) = e^{x-3}h(x-3)$ then $\mathscr{L}\left[f(x)\right](s) = \dfrac{e^{-3s}}{s-1}$.

We state the second translation property:

Theorem. If $f \in \mathcal{T}$ such that f is also integrable on $[0, +\infty)$ then for any $a \in \mathbb{R}$ it holds

$$\mathscr{L}\left[e^{ax}f(x)\right](s) = \mathscr{L}\left[f(x)\right](s-a)\,.$$

Proof. Apply the transform definition

$$\begin{aligned}\mathscr{L}\left[e^{ax}f(x)\right](s) &= \int_0^{+\infty} e^{ax}f(x)e^{-sx}\mathrm{d}x \\ &= \int_0^{+\infty} f(x)e^{-(s-a)x}\mathrm{d}x = \mathscr{L}\left[f(x)\right](s-a)\,.\end{aligned}$$

□

For instance, the Laplace transform of $f(x) = e^{-x}\sin x$ is

$$\mathscr{L}\left[e^{-x}\sin x\right](s) = \mathscr{L}\left[\sin x\right](s+1) = \frac{1}{(s+1)^2+1} = \frac{1}{s^2+2s+2}\,.$$

It is useful to study the relationship between the transformation and rescaling of the variable.

Theorem. If $f \in \mathcal{T}$ such that f is also integrable on $[0,+\infty)$ with abscissa of convergence σ_f and if $a > 0$ then

$$\mathscr{L}\left[f(ax)\right](s) = \frac{1}{a}\mathscr{L}\left[f(x)\right]\left(\frac{s}{a}\right)\,.$$

Proof. Recalling the definition of transformation, we see that after changing variable $ax = y$,

$$\mathscr{L}\left[f(ax)\right](s) = \int_0^{+\infty} f(ax)e^{-sx}\mathrm{d}x = \frac{1}{a}\int_0^{+\infty} f(y)e^{-\frac{s}{a}y}\mathrm{d}y\,.$$

From which the thesis follows. □

Since the transform is a parametric integral, making the appropriate hypotheses about its derivative is interesting.

Theorem. If $f \in \mathcal{T}$ such that f is also integrable on $[0,+\infty)$ then

$$\frac{\mathrm{d}}{\mathrm{d}s}\mathscr{L}\left[f(x)\right](s) = -\mathscr{L}\left[xf(x)\right](s)\,.$$

Proof. The derivation under the integral sign is possible due to the hypothesis of summability on f; hence

$$\begin{aligned}\frac{\mathrm{d}}{\mathrm{d}s}\mathscr{L}\left[f(x)\right](s) &= \frac{\mathrm{d}}{\mathrm{d}s}\int_0^{+\infty} e^{-sx}f(x)\mathrm{d}x = \int_0^{+\infty}\frac{\mathrm{d}}{\mathrm{d}s}e^{-sx}f(x)\mathrm{d}x \\ &= \int_0^{+\infty} e^{-sx}\left(-xf(x)\right)\mathrm{d}x\end{aligned}$$

which is exactly our statement. □

Similarly, the second and higher-order derivative formulas can be deduced.

$$\frac{\mathrm{d}^2}{\mathrm{d}s^2}\mathscr{L}\left[f(x)\right](s) = \mathscr{L}\left[x^2 f(x)\right](s), \quad \frac{\mathrm{d}^n}{\mathrm{d}s^n}\mathscr{L}\left[f(x)\right](s) = \mathscr{L}\left[(-1)^n x^n f(x)\right](s)$$

We omit the details of the proof, which are essentially the same as those just exposed. For example

$$\mathscr{L}\left[x\,\sin(ax)\right](s) = -\mathscr{L}\left[-x\,\sin(ax)\right](s) = -\frac{\mathrm{d}}{\mathrm{d}s}\mathscr{L}\left[\sin(ax)\right](s) = -\frac{\mathrm{d}}{\mathrm{d}s}\frac{a}{s^2+a^2}$$

in conclusion

$$\mathscr{L}\left[x\,\sin(ax)\right](s) = \frac{2as}{(s^2+a^2)^2}$$

Laplace transform and differential equations

Laplace transform is an injective map, as a matter of fact, when operating on continuous or piecewise continuous functions. Indeed, the natural domain of the Laplace transformation is the complex plane. In this environment, the problem of the numerical inversion of the transform can be thoroughly studied. For our purposes, which mainly concern linear ordinary differential equations, we will deal only with cases in which it is possible to invert symbolically in a way quite analogous to the square root extraction for perfect squares. This allows us to deal with the transformation in the real field. For the complete treatment extended to the complex case, we refer to the classic monograph of Doetsch [38]. The inversion of the transform is significant when, as we shall see, it is used for the solution of linear differential equations because the transformation reduces the differential problem to an algebraic problem. Still, the problem's difficulty is shifted to the subsequent inversion of the solution of the algebraic problem obtained by transformation. Laplace transform is beneficial in dealing with linear differential equations by operating on the derivatives.

Proposition 1.30. If y is an exponential order $\mathcal{C}^1$ function such that both y and y' are integrable in $[0,+\infty)$, then

$$\mathscr{L}\left[y'(x)\right](s) = s\,\mathscr{L}\left[y\right](s) - y(0)$$

and if $y \in \mathcal{C}^2$ and y'' is integrable in $[0,+\infty)$, then

$$\mathscr{L}\left[y''(x)\right](s) = s^2\,\mathscr{L}\left[y\right](s) - s\,y(0) - y'(0)\,.$$

Proof. We limit ourselves here to the first affirmation since the proof of the second is completely analogous. Integrating by parts, we find

$$\begin{aligned}\mathscr{L}\left[y'(x)\right](s) &= \lim_{b\to+\infty}\int_0^b y'(x)e^{-sx}\mathrm{d}x \\ &= \lim_{b\to+\infty}\left(\left[y(x)e^{-sx}\right]_{x=0}^{x=b} + s\int_0^b y(x)e^{-sx}\mathrm{d}x\right).\end{aligned}$$

□

Example. Consider equation

$$\begin{cases} y'' + 4y = 16x\,, \\ y(0) = 3,\ y'(0) = -6\,. \end{cases}$$

Taking the Laplace transform of both sides, writing $\mathscr{L}[y](s) = Y(s)$ we get

$$\mathscr{L}[y'' + 4y](s) = \mathscr{L}[16x](s)$$

thus using proposition 1.30 we get

$$s^2\,Y(s) - s\,y(0) - y'(0) + 4Y(s) = \frac{16}{s^2},$$

so that

$$s^2\,Y(s) - 3\,s + 6 + 4Y(s) = \frac{16}{s^2}$$

or

$$(s^2 + 4)Y(s) = 3s - 6 + \frac{16}{s^2},$$

hence we can solve for $Y(s)$ obtaining:

$$Y(s) = \frac{3s}{s^2+4} - \frac{10}{s^2+4} + \frac{4}{s^2}\,.$$

In other words, we infer that

$$\mathscr{L}[y(x)](s) = 3\mathscr{L}[\cos(2x)](s) - 5\mathscr{L}[\sin(2x)](s) + 4\mathscr{L}[x](s)\,.$$

Hence, thanks to the linearity of the transform, the solution to the given initial value problem is

$$y(x) = 3\cos(2x) - 5\sin(2x) + 4x\,.$$

It is worth noting that the formulæ for the Laplace transform of derivatives, automatically incorporates the initial values of $y(0)$ and $y'(0)$.

The following remark will be of great use in chapter 7, dedicated to Bessel functions, to analyse differential equations with variable coefficients.

Remark 1.31.

$$\begin{aligned}
\mathscr{L}[x\,y](s) &= -\frac{\mathrm{d}}{\mathrm{d}s}\mathscr{L}[y](s) = -\frac{\mathrm{d}}{\mathrm{d}s}Y(s) \\
\mathscr{L}[x\,y'](s) &= -\frac{\mathrm{d}}{\mathrm{d}s}\mathscr{L}[y'](s) = -\frac{\mathrm{d}}{\mathrm{d}s}\left(sY(s) - y(0)\right) = -\frac{\mathrm{d}}{\mathrm{d}s}\left(sY(s)\right) \\
\mathscr{L}[x\,y''](s) &= -\frac{\mathrm{d}}{\mathrm{d}s}\mathscr{L}[y''](s) = -\frac{\mathrm{d}}{\mathrm{d}s}\left(s^2Y(s) - s\,y(0) - y'(0)\right) \\
&= -\frac{\mathrm{d}}{\mathrm{d}s}\left(s^2Y(s) - sy(0)\right)
\end{aligned}$$

Laplace transform of integrals

If, with our usual notations, $\mathscr{L}[f(x)](s) = F(s)$ we can also obtain information about the Laplace transform of

$$P(x) = \int_0^x f(\xi)\,\mathrm{d}\xi$$

Theorem 1.32.

$$\mathscr{L}[P(x)](s) = \mathscr{L}\left[\int_0^x f(\xi)\,\mathrm{d}\xi\right](s) = \frac{F(s)}{s}$$

Proof. We have $P(0) = 0$ and $P'(x) = f(x)$. Using the formula for the Laplace transform of a derivative, we see that

$$\mathscr{L}[f(x)](s) = \mathscr{L}[P'(x)](s) = s\mathscr{L}[P(x)](s) - P(0) = s\mathscr{L}[P(x)](s)$$

□

Remark. Using inverse transform, we see that the Laplace transform of the integral of $f(x)$ corresponds to the division of the Laplace transform by s, i.e.

$$\int_0^x f(\xi)\,\mathrm{d}\xi = \mathscr{L}^{-1}\left[\frac{F(s)}{s}\right](x)$$

Theorem 1.33. If $\mathscr{L}[f(x)](s) = F(s)$, provided that there exits $\lim\limits_{x\to 0}\dfrac{f(x)}{x}$, then

$$\mathscr{L}\left[\frac{f(x)}{x}\right](s) = \int_s^\infty F(\sigma)\mathrm{d}\sigma \tag{†}$$

Proof. Integrating the right-hand side of (†) for $s \in (0,\infty)$ we find, recalling the definition of Laplace transform

$$\begin{aligned}\int_s^\infty F(\sigma)\mathrm{d}\sigma &= \int_s^\infty\left(\int_0^\infty e^{-\sigma x} f(x)\mathrm{d}x\right)\mathrm{d}\sigma\\ &= \int_0^\infty\left(\int_s^\infty e^{-\sigma x} f(x)\mathrm{d}\sigma\right)\mathrm{d}x\\ &= \int_0^\infty f(x)\left(\int_s^\infty e^{-\sigma x}\mathrm{d}\sigma\right)\mathrm{d}x\\ &= \int_0^\infty f(x)\frac{e^{-sx}}{x}\mathrm{d}x = \mathscr{L}\left[\frac{f(x)}{x}\right](s)\end{aligned}$$

□

Example. Taking $f(x) = \sin x$ in Theorem 1.33 we have, recalling also the sine transform $\mathscr{L}[\sin x](s) = \dfrac{1}{1+s^2}$, that

$$\mathscr{L}\left[\frac{\sin x}{x}\right](s) = \int_0^\infty \frac{\sin x}{x}e^{-sx}\mathrm{d}s = \int_s^\infty \frac{\mathrm{d}\sigma}{1+\sigma^2}$$

Thus, we obtained an interesting integration formula.

$$\int_0^\infty \frac{\sin x}{x}\, e^{-sx} \mathrm{d}s = \frac{\pi}{2} - \arctan s$$

which for $s = 0$ provides the Dirichlet integral:

$$\int_0^\infty \frac{\sin x}{x}\, \mathrm{d}s = \frac{\pi}{2}\,.$$

Since $\mathscr{L}\left[\sin x\right](s) = \dfrac{1}{1+s^2}$ and, as is well known, $\lim\limits_{x\to 0} \dfrac{\sin x}{x} = 1$ then

$$\mathscr{L}\left[\frac{\sin x}{x}\right](s) = \int_s^\infty \frac{\mathrm{d}\sigma}{1+\sigma^2} = \frac{\pi}{2} - \arctan s = \arctan\frac{1}{s}.$$

Consequently, by Theorem 1.32, we have that

$$\mathscr{L}\left[\int_0^x \frac{\sin\xi}{\xi}\, \mathrm{d}\xi\right](s) = \frac{1}{s}\arctan\frac{1}{s}\,.$$

Theorem. Under suitable assumptions regarding the convergence of the integrals, the following identity holds:

$$\int_0^\infty \mathscr{L}\left[f\right](x)\, g(x)\, \mathrm{d}x = \int_0^\infty f(s) \mathscr{L}\left[g\right](s)\, \mathrm{d}s\,. \tag{S}$$

Proof. Explicating the Laplace transform on the left-hand side, we obtain

$$\begin{aligned}
\int_0^\infty \mathscr{L}\left[f\right](x)\, g(x)\, \mathrm{d}x &= \int_0^\infty \left(\int_0^\infty e^{-xs} f(s)\, \mathrm{d}s\right) g(x)\, \mathrm{d}x \\
&= \int_0^\infty \left(\int_0^\infty e^{-xs}\, g(x)\, \mathrm{d}x\right) f(s)\, \mathrm{d}s \\
&= \int_0^\infty \mathscr{L}\left[g\right](s) f(s)\, \mathrm{d}s\,.
\end{aligned}$$

□

The following equivalent formulation of the identity (S) is very useful in calculating integrals, which are otherwise difficult to evaluate

$$\int_0^\infty f(x)\, g(x)\, \mathrm{d}x = \int_0^\infty \mathscr{L}\left[f\right](s)\, \mathscr{L}^{-1}\left[g\right](s)\, \mathrm{d}s\,. \tag{M}$$

Example. We use (M) to compute a Frullani integral: if $a,\, b > 0$, we evaluate

$$I(a,b) = \int_0^\infty \frac{e^{-ax} - e^{-bx}}{x}\, \mathrm{d}x\,.$$

$$f(x) = e^{-ax} - e^{-bx} \circ\!\!-\!\!\bullet \mathscr{L}[f](s) = \frac{1}{a+s} - \frac{1}{b+s},$$

$$g(x) = \frac{1}{x} \circ\!\!-\!\!\bullet \mathscr{L}^{-1}[g](s) = 1.$$

$$I(a,b) = \int_0^\infty \frac{e^{-ax} - e^{-bx}}{x}\,\mathrm{d}x = \int_0^\infty \left(\frac{1}{a+s} - \frac{1}{b+s}\right)\mathrm{d}s = \ln\frac{b}{a}.$$

Example.

$$\int_0^\infty \frac{\sin^3 x}{x^2}\,\mathrm{d}x$$

Using $\sin(3x) = 3\sin x - 4\sin^3 x$ we have

$$\mathscr{L}\left[\sin^3 x\right](s) = \frac{3}{4}\mathscr{L}[\sin x](s) - \frac{1}{4}\mathscr{L}[\sin(3x)](s) = \frac{3}{4}\frac{1}{1+s^2} - \frac{1}{4}\frac{3}{9+s^2}$$

$$\mathscr{L}^{-1}\left[\frac{1}{x^2}\right](s) = s$$

Hence

$$\int_0^\infty \frac{\sin^3 x}{x^2}\,\mathrm{d}x = \frac{3}{4}\int_0^\infty \left(\frac{s}{1+s^2} - \frac{s}{9+s^2}\right)\mathrm{d}s = \frac{3}{4}\ln 3.$$

Chapter 2

Gamma and beta functions

The material presented here is based on many excellent contributions; we mention, among others [4, 6, 14, 39, 50, 55, 63, 79, 80, 85, 87, 114, 115]. Gamma and beta functions are defined in terms of parametric integrals called Euler-Legendre integrals of the second and first kind, respectively, as a reminder that the earliest contributions are due to Euler and Legendre. Notably, all the most eminent mathematicians since the eighteenth century have contributed to developing the theory of these functions. Parametric integrals that define Eulerian functions are relevant in applications, starting from probability theory and indispensable in studying all the other special functions we shall present in the following chapters. It is important to underline that, for our discussion, we have chosen to present the theory of the gamma function, respecting its historical development, which arises from the Eulerian solution of interpolating the factorial function. Vice versa, some illustrious contributions, for instance, [4, 114], choose to define the gamma function through its representation in terms of infinite product, which dates back to the subsequent contributions of Weierstrass: essentially, what for us are the Theorem 2.19 and the corollary 2.20 becomes the definition of gamma and the integral representation (2.2) a theorem. The substance does not change; they are equivalent representations of the same function.

Unlike many textbooks on special functions, the space here devoted to the two main Eulerian functions is very large, with many historical references and, above all, the treatment is self-contained. The results of Laplace and Cauchy presented in Section 2.12 rarely find space in recently published texts.

2.1 Gamma: the factorial extension

The function is known as the gamma function was introduced by Leonhard Euler (1707–1783) willing to extend the factorial to the positive real line. Euler was stimulated to tackle this problem of prolongation by Daniel Bernoulli (1700–1782) and Christian Goldbach (1690–1764). The remarkable essay by Dutka [42] reconstructs the phases of the mathematical process that led to its definition. The "birth" of the gamma function can be traced in two letters that Euler wrote to Goldbach between the end of 1729 and the beginning

DOI: 10.1201/9781003171546-2

of 1730. Euler introduced for $x > 0$, function $e(x)$, defined by formula (2.1) below:

$$e(x) = \int_0^1 \left(\ln \frac{1}{t} \right)^x \, dt. \tag{2.1}$$

Later, Adrien Marie Legendre (1752–1833) introduced in 1809 the $\Gamma(x)$ notation being x again a positive real number:

$$\Gamma(x) = \int_0^\infty e^{-t}\, t^{x-1} \, dt. \tag{2.2}$$

Observe that from (2.1) and (2.2) follows $\Gamma(x+1) = e(x)$. Gamma, indeed, prolongs the factorial. This is the meaning of the following theorem.

Theorem. If $n \in \mathbb{N}$ then

$$\Gamma(n+1) = n!. \tag{2.3}$$

Moreover for any $x > 0$ gamma verifies:

$$\Gamma(x+1) = x\,\Gamma(x). \tag{2.4}$$

Proof. Formula (2.4) follows integrating by parts (2.2) when x is replaced by $x+1$:

$$\Gamma(x+1) = \int_0^\infty t^{x}\, e^{-t} \, dt = \left[-t^{x}\, e^{-t} \right]_0^\infty + x \int_0^\infty t^{x-1}\, e^{-t} \, dt = x\,\Gamma(x)\,.$$

When the argument is a positive integer from (2.2), the following applies:

$$\Gamma(1) = \int_0^\infty e^{-t} dt = 1.$$

Using (2.4) we get:

$$\Gamma(2) = 1 \cdot \Gamma(1) = 1, \quad \Gamma(3) = 2 \cdot \Gamma(2) = 2, \quad \Gamma(4) = 3 \cdot \Gamma(3) = 3 \cdot 2 = 6,$$

and so on. The formula (2.3) can be inferred by an inductive argument. □

Theorem. For $x > 0$, $\Gamma(x)$ is a $\mathcal{C}^\infty$ function.

Proof. Indeed, since it is possible to differentiate under the integral sign for Theorem 1.19, we can exchange differentiation and integration, obtaining:

$$\Gamma'(x) = \int_0^\infty e^{-t}\, t^{x-1} \, \ln t \, dt, \tag{2.5a}$$

$$\Gamma^{(2)}(x) = \int_0^\infty e^{-t}\, t^{x-1} \, (\ln t)^2 \, dt, \tag{2.5b}$$

and more generally

$$\Gamma^{(n)}(x) = \int_0^\infty e^{-t}\, t^{x-1} \, (\ln t)^n \, dt\,. \tag{2.5n}$$

□

Remark. Gamma is, by construction, strictly positive. Moreover, since (2.5b) shows that $\Gamma^{(2)}(x) \geq 0$, it follows that $\Gamma(x)$ is strictly convex. We thus infer, with elementary arguments, the existence of the following limits:

$$\lim_{x \to 0^+} \Gamma(x) = +\infty, \qquad \lim_{x \to \infty} \Gamma(x) = +\infty.$$

First, observe that

$$\Gamma(x) > \int_0^1 t^{x-1}\, e^{-t}\, \mathrm{d}t > \frac{1}{e} \int_0^1 t^{x-1}\, \mathrm{d}t = \frac{1}{x\, e}.$$

For the second, since the limit indeed exists, we can restrict natural numbers so that

$$\ell_\infty = \lim_{n \to \infty} \Gamma(n) = \lim_{n \to \infty} (n-1)! = +\infty\,.$$

Remark. Observing that $\Gamma(2) = \Gamma(1)$ and using Rolle theorem, we see that there exists $\xi \in]1, 2[$ such that $\Gamma'(\xi) = 0$. On the other hand, since $\Gamma^{(2)}(x) > 0$, there is a unique ξ such that $\Gamma'(\xi) = 0$. This means that ξ represents the absolute minimum for Γ, when $x \in]0, \infty[$. The numerical determination of ξ and $\Gamma(\xi)$ is due to Legendre and Gauss (1777–1855) and it is $\xi \simeq 1.46163$, $\Gamma(\xi) \simeq 0.88560$.

A notable property is that $\ln \Gamma(x)$ is also a convex function.

Theorem 2.1. $\Gamma(x)$ is logarithmically convex.

Proof. Recall the Bunyakovsky-Schwarz inequality for functions whose second power is summable:

$$\left(\int_0^\infty f(t)\, g(t) \mathrm{d}t \right)^2 \leq \int_0^\infty f^2(t)\, \mathrm{d}t \cdot \int_0^\infty g^2(t)\, \mathrm{d}t\,.$$

If we take $f(t) = e^{-\frac{t}{2}} t^{\frac{x-1}{2}}$, $g(t) = e^{-\frac{t}{2}} t^{\frac{x-1}{2}} \ln t$ remember the rules for the derivation of Gamma, we find the inequality:

$$\left(\Gamma'(x)\right)^2 \leq \Gamma(x)\, \Gamma^{(2)}(x).$$

Hence, we conclude that

$$\frac{\mathrm{d}^2}{\mathrm{d}x^2} \ln\left(\Gamma(x)\right) = \frac{\Gamma(x)\, \Gamma^{(2)}(x) - \left(\Gamma'(x)\right)^2}{\Gamma^2(x)} \geq 0\,.$$

□

From Theorem 2.1, we deduce a recent result due to Gautschi [53].

Theorem. For any $x > 0$ and $s \in (0, 1)$ the following inequality holds:

$$x^{1-s} \leq \frac{\Gamma(x+1)}{\Gamma(x+s)} \leq (x+1)^{1-s}. \tag{2.6}$$

Proof. For Theorem 2.1, we see that $\ln\Gamma$ is a convex function. Thus, for $t \in (0,1)$ and u, $v > 0$ we have

$$\Gamma(tu + (1-t)v) \leq \Gamma^t(u)\,\Gamma^{1-t}(v)\,.$$

Choosing $u = x$, $v = x+1$ and $t = 1-s$ we obtain

$$\Gamma(x+s) \leq \Gamma^{1-s}(x)\,\Gamma^s(x+1) = x^{s-1}\,\Gamma(x+1). \tag{2.6a}$$

While, choosing $u = x+s, v = x+s+1$ and $t = s$ we have

$$\Gamma(x+1) \leq \Gamma^s(x+s)\,\Gamma^{1-s}(x+s+1) = (x+s)^{1-s}\Gamma(x+s). \tag{2.6b}$$

Inequality (2.6) follows by multiplying the reciprocal of (2.6a) with (2.6b). □

Gamma related integrals

It is possible to express a wider family of integrals using gamma; this section provides the most significant. We warm up with a straightforward consequence of the definition of gamma.

Proposition 2.2. If $z > 0$ and $x > 0$ then

$$\int_0^\infty t^{x-1}e^{-z\,t}\mathrm{d}t = \frac{1}{z^x}\,\Gamma(x)\,. \tag{2.7}$$

Proof. Formula (2.7) follows from (2.2) changing the variable as $s = z\,t$. □

Remark. Proposition 2.2 extends to the case in which z is a complex number $z = a + i\,b$ with $a > 0$. This allows for explicit computation of a family of parametric integrals, which we will employ later in this chapter, formula (2.96) and (2.97) and in chapter 4, Theorem 4.5.

Proposition 2.3. Let $a > 0$, $b \in \mathbb{R}$ and $x > 0$. Then

$$\int_0^\infty e^{-at}t^{x-1}\cos(b\,t)\,\mathrm{d}t = \frac{\Gamma(x)}{(a^2+b^2)^{\frac{x}{2}}}\,\cos\left(x\,\arctan\frac{b}{a}\right), \tag{2.8a}$$

$$\int_0^\infty e^{-at}t^{x-1}\sin(b\,t)\,\mathrm{d}t = \frac{\Gamma(x)}{(a^2+b^2)^{\frac{x}{2}}}\,\sin\left(x\,\arctan\frac{b}{a}\right). \tag{2.8b}$$

Proof. If a and b are as in the hypothesis, we represent the complex number $z = a + i\,b$ in polar form:

$$z = a + i\,b = \left(a^2+b^2\right)^{\frac{1}{2}}\,e^{i\,\arctan\frac{b}{a}}\,. \tag{2.9}$$

We can express (2.7) as follows, writing I for the relevant integral:

$$I = \int_0^\infty e^{-z\,t}t^{x-1}\mathrm{d}t = \int_0^\infty e^{-a\,t}\big(\cos(b\,t) - i\,\sin(b\,t)\big)t^{x-1}\mathrm{d}t = \frac{\Gamma(x)}{(a+i\,b)^x}\,.$$

Recalling (2.9) we obtain

$$z^x = (a + i\,b)^x = \left(a^2 + b^2\right)^{\frac{x}{2}}\, e^{i\,x\,\arctan\frac{b}{a}}.$$

Plugging this representation of z, we arrive at:

$$I = \frac{\Gamma(x)}{(a^2+b^2)^{\frac{x}{2}}}\left(\cos\left(x\,\arctan\frac{b}{a}\right) - i\,\sin\left(x\,\arctan\frac{b}{a}\right)\right).$$

The thesis follows by comparing I's real and imaginary parts, giving both (2.8b) and (2.8a).

□

Corollary. If $x > 0$ and $b > 0$ then

$$\int_0^\infty t^{x-1}\cos(b\,t)\,\mathrm{d}t = \frac{1}{b^x}\,\Gamma(x)\,\cos\frac{\pi\,x}{2}, \tag{2.10a}$$

$$\int_0^\infty t^{x-1}\sin(b\,t)\,\mathrm{d}t = \frac{1}{b^x}\,\Gamma(x)\,\sin\frac{\pi\,x}{2}. \tag{2.10b}$$

Moreover if $-\pi/2 < a < \pi/2$ then

$$\int_0^\infty e^{-b\,t\,\cos a}t^{x-1}\cos\left(b\,t\,\sin a\right)\,\mathrm{d}t = b^{-x}\Gamma(x)\cos(a\,x), \tag{2.11a}$$

$$\int_0^\infty e^{-b\,t\,\cos a}t^{x-1}\sin\left(b\,t\,\sin a\right)\,\mathrm{d}t = b^{-x}\Gamma(x)\sin(a\,x). \tag{2.11b}$$

Proof. Taking the limit for $a \to 0^+$ in two equalities (2.8), we immediately get (2.10a) and (2.10b). Due to the occurrence of the term a^2+b^2 in formulas (2.8), replacing a with $b\,\cos a$ and b with $b\,\sin a$ in (2.8a) and (2.8b) respectively we obtain (2.11a) and (2.11b).

□

2.2 Ascending factorials, binomial coefficients

The functional relation (2.4) can be iterated, so if $n \in \mathbb{N}$ and $x > 0$:

$$\Gamma(x+n) = (x+n-1)\,(x+n-2)\,\cdots\,(x+1)\,x\,\Gamma(x). \tag{2.12}$$

We are thus naturally led to the following

Definition 2.4. The ratio

$$(x)_n := \frac{\Gamma(x+n)}{\Gamma(x)} = (x+n-1)\,(x+n-2)\,\cdots\,(x+1)\,x \tag{2.13}$$

is called the Pochhammer symbol or ascending factorial of $x > 0$.

Remark. Looking on the right-hand side of (2.13), we acknowledge that $(x)_n$ can be defined for $x \in \mathbb{C}$. The case of $x = -m$ where m is a nonnegative integer is peculiar. We will examine it later in proposition 2.7.

The notation $(x)_n$ was introduced by Pochhammer (1841–1920). It is important to know that this notation is only sometimes adopted. For example, Carlson's monograph, [29] uses the notation (x, n) while in enumerative combinatorics are used $x^{(n)}$ or $x^{\overline{n}}$, see [71].

Remark.

$$(1)_n = n!, \quad (2)_n = (n+1)!, \quad (3)_n = \frac{(n+2)!}{2}, \quad (4)_n = \frac{(n+3)!}{3!}. \tag{2.14}$$

Hence, by an inductive argument, if $m \in \mathbb{N}$, we have

$$(m+1)_n = \frac{(n+m)!}{m!}. \tag{2.15}$$

Moreover, we also have:

$$(x)_{n+1} = (x+n)\,(x)_n.$$

The following identity will be helpful in the theory of hypergeometric functions, specifically chapter 9, proposition 9.2.

Proposition 2.5.

$$(x)_{n+1} = x\,(x+1)_n. \tag{2.16}$$

Proof. Let us compute the left-hand side of (2.16);

$$(x)_{n+1} = \frac{\Gamma(x+n+1)}{\Gamma(x)} = \frac{(x+n)\Gamma(x+n)}{\Gamma(x)}. \tag{2.17}$$

Then, looking on the right-hand side of (2.16), we evaluate

$$(x+1)_n = \frac{\Gamma(x+n+1)}{\Gamma(x+1)} = \frac{(x+n)\Gamma(x+n)}{x\Gamma(x)}.$$

Then solve for $x + n$, obtaining

$$x + n = \frac{(x+1)_n\, x\Gamma(x)}{\Gamma(x+n)}.$$

Now inserting in (2.17), formula (2.16) holds. □

Ascending factorials are also linked nicely to double factorials:

$$(2n)!! = 2^n (1)_n, \quad (2n+1)!! = 2^{n+1} \left(\tfrac{1}{2}\right)_{n+1}.$$

From here, we can see how it is possible to represent triples, quadruple factorials, etc., using Pochhammer symbols. For example, in the case of triple factorials, we have

$$(3n)!!! = 3^n (1)_n, \quad (3n+1)!!! = 3^{n+1} \left(\tfrac{1}{3}\right)_{n+1}, \quad (3n+2)!!! = 3^{n+1} \left(\tfrac{2}{3}\right)_{n+1}.$$

We demonstrate the duplication property of ascending factorials.

Theorem 2.6. If $x \in \mathbb{R}$ and $n \in \mathbb{N}$ then

$$(x)_{2n} = 2^{2n} \left(\frac{x}{2}\right)_n \left(\frac{x}{2} + \frac{1}{2}\right)_n . \tag{2.18}$$

Proof. The proof is obtained by direct computation of the symbol after reorganising the $2n$ factor according to their parity:

$$\begin{aligned}(x)_{2n} &= x(x+1)(x+2)(x+3)(x+4)\cdots(x+2n-2)(x+2n-1)\\ &= [x(x+2)(x+4)\cdots(x+2n-2)]\,[(x+1)(x+3)\cdots(x+2n-1)]\\ &= 2^{2n}\left[\tfrac{x}{2}(\tfrac{x}{2}+1)(\tfrac{x}{2}+2)\cdots(\tfrac{x}{2}+n-1)\right]\left[(\tfrac{x}{2}+\tfrac{1}{2})(\tfrac{x}{2}+\tfrac{3}{2})\cdots(\tfrac{x}{2}+\tfrac{n}{2}-1)\right]\\ &= 2^{2n}\left(\tfrac{x}{2}\right)_n\left(\tfrac{x}{2}+\tfrac{1}{2}\right)_n .\end{aligned}$$

□

Remark. Also, formula (2.18) given in Theorem 2.6, makes sense for $x \in \mathbb{C}$. This fact preludes to the point, which we will examine in detail, that gamma can be extended to complex values except for negative integers.

The Pochhammer symbol has a particular behaviour when it is evaluated in correspondence with negative integers.

Proposition 2.7. If n, $m \in \mathbb{N}$, then

$$(-m)_n = \begin{cases} (-1)^n \dfrac{m!}{(m-n)!} & \text{if } 1 \le n \le m \\ 0 & \text{if } n \ge m+1. \end{cases} \tag{2.19}$$

Or, in terms of binomial coefficients:

$$(-m)_n = \begin{cases} (-1)^n n! \dbinom{m}{n} & \text{if } 1 \le n \le m \\ 0 & \text{if } n \ge m+1. \end{cases} \tag{2.19a}$$

It is worth noting that the connections between gamma functions and binomial coefficients are deep: an important branch in Mathematics we address to the monograph [108] for extensive and sound treatment of this topic.

The following identity will be used in the study of hypergeometric series; see chapters 9, 10.

Remark. For p, q positive integers

$$\frac{\left(1+\frac{p}{q}\right)_n}{\left(\frac{p}{q}\right)_n} = 1 + \frac{n\,q}{p}.$$

The ascending factorial is involved in extending the notion of binomial coefficient when the "numerator" is not an integer. In fact when $\alpha \in \mathbb{C}$ we have

$$\binom{\alpha}{n} = \frac{\alpha(\alpha-1)\dots(\alpha-(n-1))}{n!}. \tag{2.20}$$

The numerator in (2.20) is not an ascending factorial; on the contrary, it is indeed the definition of a descending factorial, which we will represent with the notation $x^{(n)}$. We formalise this fact in the following:

Definition. The descending factorial of $\alpha \in \mathbb{C}$ is given by

$$\alpha^{(n)} = \alpha(\alpha-1)\dots(\alpha-(n-1)).$$

Remark. Notice that $n^{(n)} = n!$. Also, we can express the descending factorial using gamma as

$$\alpha^{(n)} = \frac{\Gamma(\alpha+1)}{\Gamma(\alpha+1-n)}.$$

It follows that ascending and descending factorial are linked by identity

$$\alpha^{(n)} = (-1)^n(-\alpha)_n\,. \tag{2.21}$$

Henceforth, we can express the binomial series as

$$(1+x)^\alpha = \sum_{n=0}^{\infty}\binom{\alpha}{n}x^n = \sum_{n=0}^{\infty}\frac{\alpha^{(n)}}{n!}x^n \tag{2.22}$$

The meaning of the identity (2.21) becomes clear when considering the power series expansion of $(1+x)^{-\alpha}$

$$(1+x)^{-\alpha} = \sum_{n=0}^{\infty}(-1)^n\frac{(\alpha)_n}{n!}x^n$$

which involves the ascending factorial.

Another interesting couple of identities related to Pochhamer's symbols is given the following:

Theorem 2.8.

$$\Gamma\left(n+\tfrac{1}{2}\right) = \left(n-\tfrac{1}{2}\right)\left(n-\tfrac{3}{2}\right)\cdots\tfrac{3}{2}\cdot\tfrac{1}{2}\Gamma\left(\tfrac{1}{2}\right) = \frac{(2n)!}{2^{2n}n!}\sqrt{\pi}. \tag{2.23}$$

$$\Gamma\left(n-\frac{1}{2}\right) = \frac{\sqrt{\pi}}{2^{n-1}}\prod_{k=1}^{n-1}(2k-1) = \frac{\sqrt{\pi}}{2^{n-1}}\,(2n-3)!!\,. \tag{2.24}$$

Proof. Multiplying, we find

$$\left(n-\tfrac{1}{2}\right)\left(n-\tfrac{3}{2}\right)\cdots\tfrac{3}{2}\cdot\tfrac{1}{2}=\frac{(2n-1)(2n-3)\cdots 1}{2^n}=\frac{(2n-1)!!}{2^n},$$

so that (2.23) follows recalling:

$$(2n-1)!!=\frac{(2n-1)!}{2^{n-1}(n-1)!}.$$

Finally, (2.24) follows, shifting the argument. □

2.3 Continuation of gamma to negative values

Considering the relation defining the ascending factorial from another point of view, we can extend the gamma domain to the negative reals, excluding the integers. In fact, if we rewrite (2.12) as:

$$\Gamma(x)=\frac{\Gamma(x+n)}{(x+n-1)\,(x+n-2)\,\cdots\,(x+1)\,x},$$

we can evaluate $\Gamma(x)$ also for negative, non integer, values of x. For instance, when $x\in(-1\,,0)$, gamma is given by

$$\Gamma(x)=\frac{\Gamma(x+1)}{x}.$$

In particular, we have

$$\Gamma\left(-\tfrac{1}{2}\right)=-2\,\Gamma\left(\tfrac{1}{2}\right).$$

When $x\in]-2,-1[$, the evaluation is

$$\Gamma(x)=\frac{\Gamma(2-x)}{(x+1)\,x}.$$

Thus, in particular:

$$\Gamma\left(-\tfrac{3}{2}\right)=\tfrac{4}{3}\,\Gamma\left(\tfrac{1}{2}\right).$$

In other words, $\Gamma(x)$ is defined on the real line, except at its poles $x=0,-1,-2,\cdots$, and so on, as shown in Figure 2.1.

Notice also that using Pochammer symbols, recalling again $\Gamma(\frac{1}{2})=\sqrt{\pi}$, we deduce the evaluation of gamma for the negative semi-integer arguments.

$$\begin{aligned}\Gamma\left(-n-\tfrac{1}{2}\right)&=\frac{\Gamma\left(\frac{1}{2}\right)}{\left(-n-\frac{1}{2}\right)\left(-n+\frac{1}{2}\right)\cdots\left(-n-\frac{1}{2}+n\right)}\\&=(-1)^n\,\frac{2^{2n+1}n!}{(2n+1)!}\,\sqrt{\pi}.\end{aligned}$$

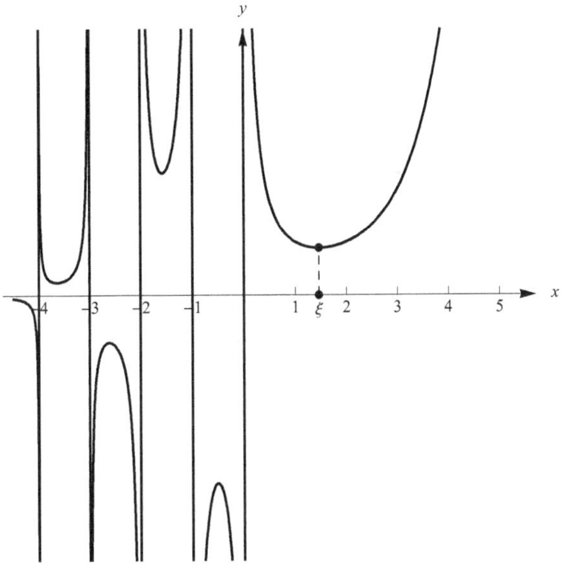

Figure 2.1: Plot of $\Gamma(x)$.

2.4 Beta

Definition. Beta function $\mathrm{B}(x, y)$, also called *Eulerian integral of the first kind*, is defined as

$$\mathrm{B}(x, y) = \int_0^1 t^{x-1}\,(1-t)^{y-1}\,\mathrm{d}t, \quad x,\, y > 0. \tag{2.25}$$

Remark. The change of variable $t = 1 - s$ provides the symmetry relation $\mathrm{B}(x, y) = \mathrm{B}(y, x)$.

Remark. An alternative definition of beta comes from the change of variable $t = \cos^2 \vartheta$, which leads to

$$\mathrm{B}(x, y) = 2 \int_0^{\frac{\pi}{2}} (\cos \vartheta)^{2x-1}\,(\sin \vartheta)^{2y-1}\,\mathrm{d}\vartheta. \tag{2.25a}$$

The change of variable $s = t(1-t)^{-1}$, provides a further alternative representation for the Beta function:

$$\mathrm{B}(x, y) = \int_0^{\infty} \frac{s^{x-1}}{(1+s)^{x+y}}\,\mathrm{d}s\,. \tag{2.26}$$

In fact, changing variable by $s = t\,(1-t)^{-1}$, we obtain

$$
\begin{aligned}
\mathrm{B}(x,y) &= \int_0^1 t^{x-1}(1-t)^{y-1}\,\mathrm{d}t = \int_0^\infty s^{x-1}(1+s)^{-x+1}(1+s)^{-y+1}(1+s)^{-2}\,\mathrm{d}s \\
&= \int_0^\infty \frac{s^{x-1}}{(1+s)^{x+y}}\,\mathrm{d}s.
\end{aligned}
$$

Moreover, from (2.26) follows a third alternative representation of Beta, which is reported in [58, page 318 entry 3.216.1]:

$$
\mathrm{B}(x,y) = \int_0^1 \frac{t^{x-1}+t^{y-1}}{(1+t)^{x+y}}\,\mathrm{d}t. \tag{2.27}
$$

If we denote by X the value of the integral on the right-hand side in (2.27), changing the variable, placing $t = \frac{1}{s}$ we see that

$$
X = \int_1^\infty \frac{s^{x-1}+s^{y-1}}{(1+s)^{x+y}}\,\mathrm{d}s.
$$

So, summing the integrals obtained and writing them both with the same integration variable, we see that

$$
\begin{aligned}
2X &= \int_0^1 \frac{t^{x-1}+t^{y-1}}{(1+t)^{x+y}}\,\mathrm{d}t + \int_1^\infty \frac{t^{x-1}+t^{y-1}}{(1+t)^{x+y}}\,\mathrm{d}t = \int_0^\infty \frac{t^{x-1}+t^{y-1}}{(1+t)^{x+y}}\,\mathrm{d}t \\
&= \int_0^\infty \frac{t^{x-1}}{(1+t)^{x+y}}\,\mathrm{d}t + \int_0^\infty \frac{t^{y-1}}{(1+t)^{x+y}}\,\mathrm{d}t = \mathrm{B}(x,y)+\mathrm{B}(y,x) = 2\mathrm{B}(x,y)
\end{aligned}
$$

proving (2.27). An alternative demonstration of (2.27) is provided in [88, pp. 18–19].

The main property of beta is its relationship with the gamma, as expressed by the Theorem 2.9 below.

Theorem 2.9. If $x,\ y > 0$, then:

$$
\mathrm{B}(x,y) = \frac{\Gamma(x)\,\Gamma(y)}{\Gamma(x+y)}. \tag{2.28}
$$

Proof. From the usual definition (2.2) for the gamma function, after the change of variable $t = u^2$, we have

$$
\Gamma(x) = 2\int_0^{+\infty} u^{2\,x-1}\,e^{-u^2}\,\mathrm{d}u\,.
$$

In the same way

$$
\Gamma(y) = 2\int_0^{+\infty} v^{2\,y-1}\,e^{-v^2}\,\mathrm{d}v.
$$

Now, we form the product of the last two integrals above, and we use the Fubini theorem to obtain the following:

$$\Gamma(x)\,\Gamma(y) = 4 \iint_{[0,+\infty)\times[0,+\infty)} u^{2x-1}\, v^{2y-1}\, e^{-(u^2+v^2)}\, \mathrm{d}u\, \mathrm{d}v.$$

At this point, we change a variable in the double integral, using polar coordinates $u = \rho\cos\vartheta$, $v = \rho\sin\vartheta$ providing:

$$\begin{aligned}\Gamma(x)\,\Gamma(y) &= 4\left(\int_0^{+\infty} \rho^{2x+2y-1} e^{-\rho^2}\,\mathrm{d}\rho\right)\left(\int_0^{\frac{\pi}{2}} \cos^{2x-1}(\vartheta)\sin^{2y-1}(\vartheta)\mathrm{d}\vartheta\right)\\ &= \Gamma(x+y)\int_0^{\frac{\pi}{2}} 2\,\cos^{2x-1}(\vartheta)\,\sin^{2y-1}(\vartheta)\,\mathrm{d}\vartheta.\end{aligned}$$

The thesis follows from (2.25a). □

Remark. Theorem 2.9 can also be shown by exchanging again the order of integration, starting from

$$\Gamma(x)\,\Gamma(y) = \int_0^\infty\int_0^\infty e^{-(t+s)}\, t^{x-1}\, s^{y-1}\,\mathrm{d}t\,\mathrm{d}s,$$

and using the change of variable $t = u\,v$ and $s = u\,(1-v)$.

The relation (2.28) provides several interesting representations of the binomial coefficients: the first is, in some sense, straightforward, and we follow [108, p. 77].

Corollary. If n, $m \in \mathbb{N}$ are such that $n - m + 1 > 0$ then:

$$\frac{1}{(n+1)\dbinom{n}{m}} = \int_0^1 x^m(1-x)^{n-m}\mathrm{d}x = \mathrm{B}(n-m+1, m+1). \tag{2.29}$$

Proof. We express the binomial coefficient in terms of gamma:

$$\begin{aligned}\frac{1}{\dbinom{n}{m}} &= \frac{\Gamma(m+1)\,\Gamma(n-m+1)}{\Gamma(n+1)}\\ &= \frac{(n+1)\,\Gamma(m+1)\,\Gamma(n-m+1)}{(n+1)\,\Gamma(n+1)} &&\text{(a)}\\ &= \frac{(n+1)\,\Gamma(m+1)\,\Gamma(n-m+1)}{\Gamma(n+2)} &&\text{(b)}\\ &= (n+1)\mathrm{B}(m+1, n-m+1) &&\text{(c)}\end{aligned}$$

In (a), we multiplied the numerator and denominator by $(n+1)$, in (b) the gamma functional property (2.4) in the denominator and (c) the relation (2.28). In conclusion, (2.29) follows from the beta definition (2.25). □

Theorem 2.9 connects beta to the central binomial coefficients.

Corollary. If $n \in \mathbb{N}$ then

$$\mathrm{B}(n,n) = \frac{2}{n\dbinom{2n}{n}}, \qquad \mathrm{B}(2n+1,n+1) = \frac{1}{(3n+1)\dbinom{3n}{n}}$$

Proof. If $n \in \mathbb{N}$ form (2.28) we see that

$$\mathrm{B}(n,n) = \frac{\Gamma^2(n)}{\Gamma(2n)} = \frac{[(n-1)!]^2}{(2n-1)!} = \frac{(n!)^2}{(2n)!}\,\frac{2n}{n^2}$$

The first statement follows, recalling that.

$$\binom{2n}{n} = \frac{(2n)!}{(n!)^2}$$

The proof of the second statement is similar, so we skip it. □

Remark. From Theorem 2.9, we also infer the following properties of beta:

$$B(x,y) = B(x,y+1) + B(x+1,y)$$
$$B(x+1,y) = B(x,y)\,\frac{x}{x+y}$$
$$B(x,y+1) = B(x,y)\,\frac{y}{x+y}$$
$$B(x,1) = \frac{1}{x}$$
$$\mathrm{B}(x_1,x_2)\,\mathrm{B}(x_1+x_2,x_3)\,\mathrm{B}(x_1+x_2+x_3,x_4) = \frac{\Gamma(x_1)\,\Gamma(x_2)\,\Gamma(x_3)\,\Gamma(x_4)}{\Gamma(x_1+x_2+x_3+x_4)}$$

Remark. Using (2.28), we can evaluate $\Gamma(\frac{1}{2})$ and, then, the probability integral. Indeed, by taking $x = z$ and $y = 1 - z$, where $0 < z < 1$, we obtain

$$\Gamma(z)\,\Gamma(1-z) = \mathrm{B}(z,1-z) = \int_0^1 t^{z-1}(1-t)^{-z}\,\mathrm{d}t = \int_0^1 \left(\frac{t}{1-t}\right)^z \frac{1}{t}\,\mathrm{d}t.$$

Now, the change of variable $y = t\,(1-t)^{-1}$, leads to

$$\Gamma(z)\;\Gamma(1-z) = \int_0^\infty \frac{y^{z-1}}{1+y}\,\mathrm{d}y. \tag{2.30}$$

In particular, the choice $z = \frac{1}{2}$ yields

$$\Gamma(\tfrac{1}{2}) = \sqrt{\pi}. \tag{2.31}$$

To see it, observe that (2.30) implies:

$$\Gamma^2\left(\frac{1}{2}\right) = \int_0^\infty \frac{1}{(1+y)\,\sqrt{y}}\,dy\,,$$

where the right-side–hand side integral is computed by setting $y = x^2$, So that

$$\Gamma^2(\tfrac{1}{2}) = 2\int_0^\infty \frac{1}{1+x^2}\,\mathrm{d}x = 2\lim_{b\to\infty} \arctan b = \pi.$$

Employing (2.31), it is now possible to evaluate the probability integral. Taking $x = \frac{1}{2}$ in (2.2) and then setting $t = x^2$, we find

$$\Gamma(\tfrac{1}{2}) = \int_0^\infty \frac{e^{-t}}{\sqrt{t}}\,\mathrm{d}t = 2\int_0^\infty e^{-x^2}\,\mathrm{d}x = \sqrt{\pi}. \tag{2.31b}$$

Finally, we can state a representation theorem which describes the family of integrals related to the Beta function.

Theorem. If x, $y > 0$ and $a < b$, then

$$\int_a^b (s-a)^{x-1}\,(b-s)^{y-1}\,\mathrm{d}s = (b-a)^{x+y-1}\,\mathrm{B}(x,y).$$

Proof. In (2.25), employ the change of variable $t = \frac{s-a}{b-a}$, to obtain

$$\begin{aligned}\mathrm{B}(x,y) &= \int_0^1 t^{x-1}(1-t)^{y-1}\,\mathrm{d}t\\ &= \int_a^b (s-a)^{x-1}(b-a)^{-x+1}(b-s)^{y-1}(b-a)^{-y+1}(b-a)^{-1}\,\mathrm{d}s\\ &= (b-a)^{-x-y+1}\int_a^b (s-a)^{x-1}(b-s)^{y-1}\,\mathrm{d}s,\end{aligned}$$

which complements our argument. □

When in particular we take $a = -1$ and $b = 1$ we get

$$\int_{-1}^1 (1+s)^{x-1}(1-s)^{y-1}\,\mathrm{d}s = 2^{x+y-1}\mathrm{B}(x,y).$$

2.5 Euler reflection formula

Euler's reflection formula, besides being, in our opinion, one of the most aesthetically intriguing formulas in Mathematics, is applied in many statements and identities about special functions, starting with the next section when we will generalise Legendre's duplication formula (2.50). We need the sine infinite product representation (1.33) to demonstrate it. From the integral definition of gamma (2.2), it follows that, apart from positive integers and semi-integers, there are no elementary expressions for the values of the function. However, the value of certain products of "conjugates" pairs is found through the sine function.

Theorem. For any $x \in (0,1)$:

$$\Gamma(x)\,\Gamma(1-x) = \frac{\pi}{\sin(\pi\,x)}. \tag{2.32}$$

Proof. We start from the beta alternate representation (2.26):

$$\mathrm{B}(x,1-x) = \Gamma(x)\,\Gamma(1-x) = \int_0^\infty \frac{s^{x-1}}{1+s}\,\mathrm{d}s\,.$$

Then, splitting the integration between $[0,1]$ and $(1,+\infty)$ after changing variable using $s = 1/\sigma$ in the unbounded interval, we obtain

$$\Gamma(x)\,\Gamma(1-x) = \int_0^1 \frac{s^{x-1}}{1+s}\,\mathrm{d}s + \int_0^1 \frac{\sigma^{-x}}{1+\sigma}\,\mathrm{d}\sigma\,.$$

Now, recalling that

$$\frac{1}{1+s} = \sum_{n=0}^{\infty}(-1)^n s^n,$$

if we write s again in place of σ in the second integral, we get exchanging, as is, in fact, permissible, integral and series

$$\begin{aligned}\Gamma(x)\,\Gamma(1-x) &= \sum_{n=0}^{\infty}\left(\int_0^1 (-1)^n s^{x+n-1}\mathrm{d}s + \int_0^1 (-1)^n s^{n-x}\mathrm{d}s\right)\\ &= \sum_{n=0}^{\infty}\frac{(-1)^n}{x+n} + \sum_{n=0}^{\infty}\frac{(-1)^n}{n+1-x}\,.\end{aligned}$$

Now, if, in the last equality, we rewrite the first sum, isolating the zero-index term, and in the second sum, we change the index as $n+1=m$, we get

$$\Gamma(x)\,\Gamma(1-x) = \frac{1}{x} + \sum_{n=1}^{\infty}\frac{(-1)^n}{x+n} + \sum_{m=1}^{\infty}\frac{(-1)^{m-1}}{m-x}\,.$$

In conclusion of our reasoning, after using n again as the sum index in the second series, we finally arrive at

$$\Gamma(x)\,\Gamma(1-x) = \frac{1}{x} + \sum_{n=1}^{\infty}\frac{(-1)^n}{x+n} + \sum_{n=1}^{\infty}\frac{(-1)^{n-1}}{n-x}\,.$$

The very last step is to add up the two terms of the two series

$$\begin{aligned}\Gamma(x)\,\Gamma(1-x) &= \frac{1}{x} + \sum_{n=1}^{\infty}(-1)^n\left(\frac{1}{x+n} - \frac{1}{n-x}\right)\\ &= \frac{1}{x} + \sum_{n=1}^{\infty}(-1)^n\frac{-2x}{n^2-x^2}\,.\end{aligned}$$

The last expression is evaluated on the right-hand side of (1.33), see page 35, and this concludes the proof. □

Remark. Going back to integral (1.22), see page 22, that was obtained using complex analysis, if we make the change of variable $x = \ln t$, we find

$$\int_0^{+\infty} \frac{t^{a-1}}{1+t} \mathrm{d}t = \frac{\pi}{\sin(a\,\pi)}. \tag{2.33}$$

Recalling (2.30), we derive again (2.32). Vice versa from (2.32), we can evaluate integral (2.33), in fact, for $x \in (0,1)$ then

$$\mathrm{B}(x, 1-x) = \int_0^{\infty} \frac{u^{x-1}}{1+u}\, \mathrm{d}u = \frac{\pi}{\sin(\pi\, x)}.$$

Remark. It is possible to use (2.32) to establish the *cosine* reflection formula, namely:

$$\Gamma\left(\tfrac{1}{2}+p\right)\Gamma\left(\tfrac{1}{2}-p\right) = \frac{\pi}{\sin\left(\left(\frac{1}{2}+p\right)\pi\right)} = \frac{\pi}{\cos\left(p\,\pi\right)} \tag{2.34}$$

in which we must assume $p \neq n + \frac{1}{2}$, for $n = 0, 1, \ldots$. In terms of the original variable x, we also have:

$$\Gamma(x)\,\Gamma(1-x) = \frac{\pi}{\cos\left(\left(x-\frac{1}{2}\right)\pi\right)}.$$

The reflection formula (2.32), also using another famous integral, see (2.36) below, computed, not surprisingly by Euler in 1769, is the main ingredient for the calculation of the next special integral of the logarithm of gamma.

Proposition 2.10.

$$\int_0^1 \ln\left(\Gamma(x)\right)\, \mathrm{d}x = \ln\left(\sqrt{2\,\pi}\right). \tag{2.35}$$

Proof. Let us start with the Euler log-sine integral evaluated by Euler in [49]:

$$\int_0^{\frac{\pi}{2}} \ln(\sin x)\, \mathrm{d}x = -\frac{\pi}{2}\ln 2. \tag{2.36}$$

The proof of (2.36) relies on the symmetry property:

$$\int_0^a f(x)\, \mathrm{d}x = \int_0^a f(a-x)\, \mathrm{d}x. \tag{2.37}$$

That for the left-hand side of (2.36) reads as

$$\int_0^{\frac{\pi}{2}} \ln(\sin x)\, \mathrm{d}x = \int_0^{\frac{\pi}{2}} \ln\left(\sin(\tfrac{\pi}{2}-x)\right)\, \mathrm{d}x = \int_0^{\frac{\pi}{2}} \ln(\cos x)\, \mathrm{d}x.$$

Write E for the left hand side of (2.36), then

$$\begin{aligned}
2E &= \int_0^{\frac{\pi}{2}} (\ln(\sin x) + \ln(\cos x))\,\mathrm{d}x = \int_0^{\frac{\pi}{2}} \ln(\sin x\,\cos x)\,\mathrm{d}x \\
&= \int_0^{\frac{\pi}{2}} \ln\left(\tfrac{1}{2}\,\sin(2x)\right)\mathrm{d}x = \int_0^{\frac{\pi}{2}} \ln\left(\sin(2x)\right)\mathrm{d}x + \int_0^{\frac{\pi}{2}} \ln\frac{1}{2}\,\mathrm{d}x \\
&= \frac{1}{2}\int_0^{\pi} \ln(\sin u)\,\mathrm{d}u + \frac{\pi}{2}\ln\frac{1}{2} && \text{(a)} \\
&= \frac{1}{2}\left(2\int_0^{\frac{\pi}{2}} \ln(\sin u)\,\mathrm{d}u\right) - \frac{\pi}{2}\ln 2 && \text{(b)} \\
&= E - \frac{\pi}{2}\ln 2.
\end{aligned}$$

We notice that (a) depends on the change of variable $2x = u$, while (b) derives from a second symmetry property

$$\int_0^a f(x)\mathrm{d}x = 2\int_0^{\frac{a}{2}} f(x)\mathrm{d}x\,. \tag{2.38}$$

Notice that in our situation $a = \pi$ and that the integrand $f(x)$ obeys to the symmetry property e $f(a - x) = f(x)$, see [100, page 266 Corollary 10.32] or [13, page 126]. Thus we infer

$$E = -\frac{\pi}{2}\ln 2.$$

Moreover, for (2.38), we have

$$\int_0^{\pi} \ln(\sin x)\,\mathrm{d}x = 2E = -\pi\,\ln 2. \tag{2.39}$$

Now we can evaluate (2.35). We begin invoking again (2.37) to infer:

$$\int_0^1 \ln\left(\Gamma(x)\right)\mathrm{d}x = \int_0^1 \ln\left(\Gamma(1-x)\right)\mathrm{d}x.$$

In conclusion, we see that

$$\begin{aligned}
2\int_0^1 \ln\left(\Gamma(x)\right)\mathrm{d}x &= \int_0^1 \left(\ln\left(\Gamma(x)\right) + \ln\left(\Gamma(1-x)\right)\right)\mathrm{d}x \\
&= \int_0^1 \ln\left(\Gamma(x)\,\Gamma(1-x)\right)\mathrm{d}x \\
&= \int_0^1 \ln\left(\frac{\pi}{\sin(\pi x)}\right)\mathrm{d}x && \text{(c)} \\
&= \int_0^1 \ln\pi\,\mathrm{d}x - \int_0^1 \ln\left(\sin(\pi x)\right)\mathrm{d}x \\
&= \ln\pi - \frac{1}{\pi}\int_0^{\pi} \ln(\sin u)\,\mathrm{d}u && \text{(d)} \\
&= \ln\pi + \ln 2 && \text{(e)} \\
&= \ln(2\pi).
\end{aligned}$$

We observe that (c) follows from the reflection formula, (d) forms the change of variable $\pi x = u$, and (e) from (2.39); thus, (2.35) is proved. □

As a consequence of the functional property of gamma and the proposition 2.10 just proved, we have a second interesting integration formula for the logarithm of gamma, due to Raabe (1801–1859).

Proposition 2.11. If $x > 0$ then

$$\int_x^{x+1} \ln\Gamma(t)\,\mathrm{d}t = x\,(\ln x - 1) + \ln\left(\sqrt{2\,\pi}\right). \tag{2.40}$$

Proof. We start observing that

$$\begin{aligned}\int_0^x \ln\Gamma(t+1)\,\mathrm{d}t &= \int_0^x \ln\left(t\,\Gamma(t)\right)\,\mathrm{d}t = \int_0^x \ln t\,\mathrm{d}t + \int_0^x \ln\Gamma(t)\,\mathrm{d}t \\ &= x\,(\ln x - 1) + \int_0^x \ln\Gamma(t)\,\mathrm{d}t\,.\end{aligned}$$

Or, alternatively

$$\begin{aligned}\int_0^x \ln\Gamma(t)\,\mathrm{d}t &= \int_0^x \ln\Gamma(t+1)\,\mathrm{d}t - x\,(\ln x - 1) \\ &= \int_1^{x+1} \ln\Gamma(s)\,\mathrm{d}s - x\,(\ln x - 1)\,.\end{aligned} \tag{$\diamond$}$$

On the other side, since from the additivity of the integral, we have

$$\int_x^{x+1} \ln\Gamma(t)\,\mathrm{d}t = \int_0^{x+1} \ln\Gamma(t)\,\mathrm{d}t - \int_0^x \ln\Gamma(t)\,\mathrm{d}t\,.$$

And using ($\diamond$), we arrive at (returning to the integration variable t):

$$\begin{aligned}\int_x^{x+1} \ln\Gamma(t)\,\mathrm{d}t &= \int_0^{x+1} \ln\Gamma(t)\,\mathrm{d}t - \int_1^{x+1} \ln\Gamma(t)\,\mathrm{d}t + x\,(\ln x - 1) \\ &= -\left(\int_1^{x+1} \ln\Gamma(t)\,\mathrm{d}t - \int_0^{x+1} \ln\Gamma(t)\,\mathrm{d}t\right) + x\,(\ln x - 1) \\ &= \int_0^1 \ln\Gamma(t)\,\mathrm{d}t + x\,(\ln x - 1)\,.\end{aligned}$$

Thus, given (2.35), the proof of (2.40) is completed. □

Unlike the previous two methods, we conclude the section by presenting a demonstration of the reflection formula based on elementary methods. In fact, in our first proof, we employed the formula (1.33), which is a consequence of the use of the Fourier series. The second proof is based on complex analysis (residue theorem) to evaluate the integral (1.22).

To the best of my knowledge, this remarkable proof is new. Kerkeni Elies published his demonstration on his Instagram profile, @elies.calculus, and authorised me to report it here—my heartfelt thanks to him.

As often happens when using elementary methods, we must do some preparatory work. Let's start with a parametric integral, indeed due to Laplace, [73, Livre premiere, p. 99] presented, among others in [85, pages 107–110], [32] and also taken up in [101, pages 255–256], of which, for completeness of exposition, we report the interesting calculation procedure, based on the differentiation of parametric integrals. The Laplace integral is

$$L(z) = \int_0^\infty \frac{\cos(z\,t)}{1+t^2}\,\mathrm{d}t \tag{2.41}$$

where we assume $z > 0$. We first show that

$$L(z) = \frac{\pi}{2}\,e^{-z}\,. \tag{2.42}$$

Integrating (2.41) by parts, we obtain

$$z\,L(z) = 2\int_0^\infty \frac{t\,\sin(z\,t)}{(1+t^2)^2}\,\mathrm{d}t. \tag{2.41a}$$

Then, we differentiate (2.41a) with respect to z:

$$L(z) + z\,L'(z) = 2\int_0^\infty \frac{t^2\,\cos(z\,t)}{(1+t^2)^2}\,dt\,. \tag{2.41b}$$

Now, we use the partial decomposition:

$$\frac{t^2}{(1+t^2)^2} = \frac{1}{1+t^2} - \frac{1}{(1+t^2)^2},$$

and plugging it into (2.41b), we get

$$L(z) + z\,L'(z) = 2\,L(z) - 2\int_0^\infty \frac{\cos(z\,t)}{(1+t^2)^2}\,dt\,,$$

that is

$$z\,L'(z) - L(z) = -2\int_0^\infty \frac{\cos(z\,t)}{(1+t^2)^2}\,dt\,. \tag{2.41c}$$

Differentiating (2.41c) we obtain

$$z\,L''(z) = 2\int_0^\infty \frac{t\,\sin(z\,t)}{(1+t^2)^2}\,\mathrm{d}t.$$

So that is recalling (2.41a), we obtain the differential equation $L''(z) = L(z)$, hence

$$L(z) = c_1\,e^{z} + c_2\,e^{-z}.$$

To compute the integration constants c_1 and c_2 we start from (2.41), evaluated at $z = 0$ yielding immediately

$$L(0) = \int_0^\infty \frac{\mathrm{d}t}{1+t^2} = \frac{\pi}{2}.$$

Thus $c_1 + c_2 = \pi/2$. Moreover, from (2.41a), we see that

$$L(z) = \frac{2}{z} \int_0^\infty \frac{t \sin(z\,t)}{(1+t^2)^2}\,\mathrm{d}t.$$

Hence

$$\lim_{z\to+\infty} L(z) = 0\,.$$

So we infer that $c_1 = 0$, and this finally implies (2.42).

For our treatment of the reflection formula, it is useful to change the integration variable in the Laplace integral formula (2.42), setting $t^2 = u$ so that (2.42) provides our first key formula

$$e^{-z} = \frac{1}{\pi} \int_0^\infty \frac{\cos\left(\,z\,\sqrt{u}\,\right)}{\sqrt{u}\,(1+u)}\,\mathrm{d}u. \tag{2.43}$$

The second key piece is the formula (2.10a) of page 62, which we rewrite here

$$\int_0^\infty z^{s-1} \cos(\,b\,z\,)\,\mathrm{d}z = \frac{\Gamma(s)}{b^s}\,\cos\left(\frac{\pi}{2}\,s\right) \tag{2.10a}$$

where we assume $s > 0$.

Once this long preparatory phase is over, we can present the alternative proof of (2.32). We start from the basic definition of Gamma given in (2.2), assume $0 < s < 1$:

$$\begin{aligned}
1 &= \frac{1}{\Gamma(s)} \int_0^\infty z^{s-1} e^{-z}\,\mathrm{d}z \\
&= \frac{1}{\pi\,\Gamma(s)} \int_0^\infty z^{s-1} \left(\int_0^\infty \frac{\cos\left(\,z\,\sqrt{u}\,\right)}{\sqrt{u}\,(1+u)}\,\mathrm{d}u \right) \mathrm{d}z && \text{(a)}\\
&= \frac{1}{\pi\,\Gamma(s)} \int_0^\infty \frac{1}{\sqrt{u}\,(1+u)} \left(\int_0^\infty z^{s-1} \cos\left(\,z\,\sqrt{u}\,\right)\mathrm{d}z \right) \mathrm{d}u && \text{(b)}\\
&= \frac{1}{\pi\,\Gamma(s)} \int_0^\infty \frac{1}{\sqrt{u}\,(1+u)}\,\frac{\Gamma(s)}{u^{s/2}}\,\cos\left(\frac{\pi}{2}\,s\right) \mathrm{d}u && \text{(c)}\\
&= \frac{\cos\left(\frac{\pi}{2}\,s\right)}{\pi} \int_0^\infty \frac{u^{-\frac{s+1}{2}}}{1+u}\,\mathrm{d}u && \text{(d)}\\
&= \frac{\cos\left(\frac{\pi}{2}\,s\right)}{\pi}\,\mathrm{B}\left(\frac{1-s}{2}, \frac{1+s}{2}\right) && \text{(e)}\\
&= \frac{\cos\left(\frac{\pi}{2}\,s\right)}{\pi}\,\Gamma\left(\frac{1-s}{2}\right)\Gamma\left(\frac{1+s}{2}\right) && \text{(f)}
\end{aligned}$$

here (a) follows from (2.43), (b) follows from Fubini's theorem, (c) follows from (2.10a), (d) follows from straightforward simplifications, (e) follows from the Beta representation provided by the formula (2.26) and (f) from the Theorem 2.9. Observe that

$$0 < s < 1 \implies 0 < \frac{1 \pm s}{2} < 1$$

Summarising, we have demonstrated that

$$\frac{\pi}{\cos\left(\frac{\pi}{2}\, s\right)} = \Gamma\left(\frac{1-s}{2}\right)\Gamma\left(\frac{1+s}{2}\right) \tag{2.44}$$

We are finally done, since (2.32) follows from (2.44) introducing the parameter σ defined by

$$\frac{1+s}{2} = 1 - \sigma \iff s = 2\,\sigma - 1.$$

2.6 Stirling formula

The asymptotic behaviour of $n!$ was a problem of central importance in the eighteenth century studied in parallel with Eulerian integrals due to the close correlation between these questions. In 1730, the formula due to the Stirling (1692–1770) was established:

Theorem 2.12.

$$\lim_{n\to\infty} \frac{n!\, e^n}{n^n\,\sqrt{n}} = \sqrt{2\,\pi}. \tag{2.45}$$

Proof. The first step consists of the definite integral, the computation follows immediately from (2.39)

$$\int_0^1 \ln\left(\frac{\sin(\pi\, x)}{\pi\, x}\right) \mathrm{d}x = 1 - \ln(2\,\pi). \tag{2.46}$$

The second step exploits (1.41) once it has been observed that the logarithm transforms products into sums.

$$\ln\left(\frac{\sin(\pi\, x)}{\pi\, x}\right) = \ln\left(\prod_{n=1}^{\infty}\left(1 - \frac{x^2}{n^2}\right)\right) = \sum_{n=1}^{\infty} \ln\left(1 - \frac{x^2}{n^2}\right).$$

In this way, we can recalculate the integral (2.46) by integrating the series term by term and then comparing the two representations obtained:

$$\int_0^1 \ln\left(\frac{\sin(\pi\, x)}{\pi\, x}\right) \mathrm{d}x = \sum_{n=1}^{\infty}\int_0^1 \ln\left(1 - \frac{x^2}{n^2}\right) \mathrm{d}x\,. \tag{2.47}$$

Integration by series in (2.47) is legitimate because the series consists of terms of constant sign (negative in this case), so monotone convergence can be invoked. The calculation of integrals also deserves some attention because in the case $n = 1$, the point $x = 1$ is indeed a singularity, which we will remove by using an appropriate ansatz in the integration by parts:

$$\begin{aligned}\int_0^1 \ln\left(1-\frac{x^2}{n^2}\right) \mathrm{d}x &= \left[(x-1)\ln\left(1-\frac{x^2}{n^2}\right)\right]_0^1 - \int_0^1 (x-1)\frac{2x}{x^2-n^2}\,\mathrm{d}x \\ &= \int_0^1 \left(-2+\frac{n-1}{n-x}+\frac{n+1}{n+x}\right)\mathrm{d}x \\ &= -2-(n-1)\ln(n-1)+(n+1)\ln(n+1)-2\ln n.\end{aligned}$$

In the partial sum of the series on the right-hand side of (2.47), we use the calculation is just done

$$\begin{aligned}s_n &= \sum_{k=1}^{n} \int_0^1 \ln\left(1-\frac{x^2}{k^2}\right)\mathrm{d}x \\ &= \sum_{k=1}^{n} \left(-2-(k-1)\ln(k-1)+(k+1)\ln(k+1)-2\ln k\right) \\ &= -2n+n\ln n+(n+1)\ln(n+1)-2\ln(n!)\,.\end{aligned}$$

Notice that we used telescopic identity:

$$\sum_{k=1}^{n} \left[(k+1)a_{k+1}-(k-1)a_{k-1}\right] = n a_n + (n+1)a_{n+1} - a_1,$$

having observed that in our case $a_k = \ln k$ and a_1 is to be understood as:

$$\lim_{k\to 1+} (k-1)\ln(k-1) = 0\,.$$

We now rearrange the expression found for s_n as follows:

$$\begin{aligned}s_n &= -2\left[\ln(n!)+n-n\ln(n)\right]-n\ln(n)+(n+1)\ln(n+1) \\ &= -2\left[\ln(n!)+n-n\ln(n)\right]+n\ln\left(1+\frac{1}{n}\right)+\ln(n+1) \\ &= -2\left[\ln(n!)+n-n\ln(n)-\frac{1}{2}\ln(n+1)\right]+n\ln\left(1+\frac{1}{n}\right) \\ &= -2\ln\left[\frac{n!\,e^n}{n^n\sqrt{n+1}}\right]+\ln\left(1+\frac{1}{n}\right)^n.\end{aligned}$$

On the other hand, for identity (2.46) yields:

$$\lim_{n\to+\infty}\left\{-2\ln\left[\frac{n!\,e^n}{n^n\sqrt{n+1}}\right]+\ln\left(1+\frac{1}{n}\right)^n\right\} = 1-\ln(2\,\pi)\,.$$

This last equality leads to

$$\lim_{n\to+\infty} \ln \frac{n!\, e^n}{n^n \sqrt{n+1}} = \ln \sqrt{2\,\pi}.$$

In conclusion, switching to exponentials, the proof of (2.45) is completed. □

The elegance of (2.45) continues to capture the attention of the mathematical community, so much so that it is practically impossible to list the demonstrations obtained with more or less elementary methods. Without any claim to originality, and even less to present the history of how the problem was tackled, we shall remark that our is founded on the representation of the sine in the infinite product (1.41) also uses the Euler integral (2.39).

The above statement, by its very nature, makes sense for sequences. In applications, however, we will also need to consider situations where the variable is real, so we present a second result that extends the validity of the asymptotic formula (2.45), We present the argument proposed in [91].

Theorem 2.13.

$$\lim_{x\to\infty} \frac{e^x\sqrt{x}}{x^x}\,\Gamma(x) = \sqrt{2\pi} \tag{2.48}$$

Proof. Let $x > 0$. Being

$$\Gamma(x) = \int_0^\infty t^{x-1} e^{-t} \mathrm{d}t = 2\int_0^\infty u^{2x-1} e^{-u^2} \mathrm{d}u.$$

We have that

$$\frac{e^x\sqrt{x}}{x^x}\,\Gamma(x) = 2\int_0^\infty e^{x-u^2}\left(\frac{u}{\sqrt{x}}\right)^{2x-1} \mathrm{d}u.$$

We change the variable by placing $u = \sqrt{x} + v$ so that

$$\frac{e^x\sqrt{x}}{x^x}\,\Gamma(x) = 2\int_{\sqrt{x}}^\infty e^{-2v\sqrt{x}}\left(1 + \frac{v}{\sqrt{x}}\right)^{2x-1} e^{-v^2} \mathrm{d}v = 2\int_{-\infty}^\infty \varphi_x(v) e^{-v^2} \mathrm{d}v.$$

Where we have introduced the function:

$$\varphi_x(v) = \begin{cases} 0 & \text{for } v \le -\sqrt{x} \\ e^{-2v\sqrt{x}}\left(1 + \dfrac{v}{\sqrt{x}}\right)^{2x-1} & \text{for } v \ge -\sqrt{x}. \end{cases}$$

Expanding in series $\ln\left(1 + \frac{v}{\sqrt{x}}\right) = \frac{v}{\sqrt{x}} - \frac{v^2}{2x} + \cdots$ we see that for each fixed v $\ln \varphi_x(v) = -v^2 + \mathrm{O}\left(\frac{1}{\sqrt{x}}\right)$ for $x \to \infty$ hence

$$\lim_{x\to\infty} \varphi_x(v) = e^{-v^2}.$$

Then we observe that $\varphi_x(v)$ reaches a maximum for $v = -1/(2\sqrt{x})$. Hence we have:

$$\varphi_x'(v) = -\frac{\sqrt{x}e^{-2v\sqrt{x}}\left(\frac{v}{\sqrt{x}}+1\right)^{2x}(2v\sqrt{x}+1)}{(v+\sqrt{x})^2}\,.$$

So, we have the estimate:

$$\varphi_x(v) \leq e\left(1-\frac{1}{2x}\right)^{2x-1}.$$

This maximum value is bounded for $x \to \infty$ because

$$\lim_{x\to\infty}\left(1-\frac{1}{2x}\right)^{2x-1} = \frac{1}{e}.$$

To recap, we have shown that

$$\lim_{x\to\infty}\varphi_x(v)e^{-v^2} = e^{-2v^2}, \quad \varphi_x(v)e^{-v^2} = \mathrm{O}(e^{-v^2}) \quad \text{for} \quad x \to \infty\,.$$

This allows, [103, page 194], to exchange between the limit and the integral:

$$\lim_{x\to\infty}\frac{e^x\sqrt{x}}{x^x}\,\Gamma(x) = 2\int_{-\infty}^{\infty} e^{-2v^2}\mathrm{d}v = \sqrt{2\pi}.$$

□

Remark. It is possible to generalise the previous statement, arriving to

$$\Gamma(x) \simeq \frac{x^x\sqrt{2\pi}}{e^x\sqrt{x}}\left(1+\frac{1}{12x}+\frac{1}{288x^2}-\frac{139}{51840x^3}-\frac{571}{2488320x^4}+\cdots+\cdots\right).$$

We will not present the demonstration here; those who wish to explore further can refer to [86].

2.7 Legendre duplication and multiplication formulas

In this section, we will present the demonstration of the theorem, known as the Gauss multiplication Theorem, although, as pointed out by A. Aycock in his recent historical note [16, page 201], there is evidence that Euler arrived at an equivalent result.

Theorem 2.14. For each $m \in \mathbb{N}$, $x > 0$, the following formula holds:

$$n^{1/2-n\,x}(2\pi)^{(n-1)/2}\Gamma(n\,x) = \Gamma(x)\,\Gamma(x+\tfrac{1}{n})\,\Gamma(x+\tfrac{2}{n})\ldots\,\Gamma(x+\tfrac{n-1}{n})\,. \quad (2.49)$$

A.M. Legendre provided the proof in the case $n = 2$, which is our starting point. Following the method presented in [46, pages 173–175], we will examine in detail to facilitate the reader cases $n = 2$ and $n = 3$, which are illustrative of the general procedure, to arrive at (2.49).

Legendre formula expresses $\Gamma(2\,x)$ in terms of $\Gamma(x)$ and $\Gamma(x+\frac{1}{2})$.

Theorem 2.15.

$$\Gamma(x)\,\Gamma(x+\tfrac{1}{2}) = 2^{1-2x}\,\pi^{\frac{1}{2}}\,\Gamma(2x). \tag{2.50}$$

Proof. We follow the proof proposed in [46], which is based on the derivation of integrals dependent on a parameter introduced by Liouville. The product of gamma on the left-hand side of (2.50) can be rewritten using the integral representation of gamma. Since the extremes are fixed, and the two variables are split, the product of the two integrals can also be represented as a double integral as

$$\Gamma(x)\,\Gamma(x+\tfrac{1}{2}) = \int_0^\infty\int_0^\infty e^{-(t+s)}t^{x-1}s^{x-1/2}\mathrm{d}t\,\mathrm{d}s\,.$$

Now we change the variable from t to τ, leaving s unchanged, by setting $t = \frac{\tau^2}{s}$ so that, we obtain

$$\begin{aligned}\Gamma(x)\,\Gamma(x+\tfrac{1}{2}) &= \int_0^\infty\int_0^\infty e^{-(\frac{\tau^2}{s}+s)}\left(\frac{\tau^2}{s}\right)^{x-1}s^{x-1/2}\,\frac{2\tau}{s}\,\mathrm{d}\tau\,\mathrm{d}s\\ &= 2\int_0^\infty \tau^{2x-1}\left(\int_0^\infty e^{-(\frac{\tau^2}{s}+s)}s^{-1/2}\,\mathrm{d}s\right)\mathrm{d}\tau\,. \end{aligned} \tag{$*$}$$

We now turn our attention to the inner integral, which is called Liouville integral, by studying it as a function of τ:

$$I(\tau) = \int_0^\infty e^{-(\frac{\tau^2}{s}+s)}s^{-1/2}\,\mathrm{d}s\,.$$

$I(\tau)$ is differentiable concerning τ, the derivation of the integral is indeed legitimate, see Theorem 1.19 on page 25, therefore

$$I'(\tau) = -2\tau\int_0^\infty e^{-(\frac{\tau^2}{s}+s)}s^{-3/2}\,\mathrm{d}s\,.$$

Let's change once again the variable from $s\to\sigma$ by setting $\frac{\tau^2}{s} = \sigma$ obtaining:

$$I'(\tau) = -2\int_0^\infty e^{-\frac{\tau^2}{\sigma}-\sigma}\sigma^{-1/2}\mathrm{d}\sigma = -2\,I(\tau)\,.$$

$I(\tau)$ solves the separable equation $I'(\tau) = -2I(\tau)$ from which $I(\tau) = c\,e^{-2\tau}$. Since for $\tau = 0$ we have

$$I(0) = \int_0^\infty e^{-s}s^{-1/2}\,\mathrm{d}s = \Gamma(\tfrac{1}{2}) = \sqrt{\pi}.$$

We infer that

$$I(\tau) = \int_0^\infty e^{-(\frac{\tau^2}{s}+s)}\, s^{-1/2}\, \mathrm{d}s = \sqrt{\pi}\, e^{-2\tau}\,.$$

Therefore, returning to (∗) we see that

$$\begin{aligned}\Gamma(x)\,\Gamma(x+\tfrac{1}{2}) &= 2\sqrt{\pi}\int_0^\infty \tau^{2x-1}e^{-2\tau}\,\mathrm{d}\tau \\ &= \sqrt{\pi}\,2^{1-2x}\int_0^\infty e^{-t}t^{2x-1}\,\mathrm{d}t \qquad (\odot)\\ &= \sqrt{\pi}\,2^{1-2x}\,\Gamma(2x)\,.\end{aligned}$$

In ($\odot$), we changed the variable of integration $\tau \to t$ with $2\tau = t$. The proof of (2.50) is completed. □

As we shall see immediately, the demonstration technique presented here can be extended to higher-order cases, starting with triplication.

Similarly to the duplication formula, we prove the so-called triplication formula.

Theorem.

$$\Gamma(x)\,\Gamma(x+\tfrac{1}{3})\,\Gamma(x+\tfrac{2}{3}) = 3^{1-3x}\,2\,\pi\,\Gamma(3x). \tag{2.51}$$

Proof. Again, we interpret the product of the three gamma values in the left-hand side of (2.51) as a triple integral:

$$\Gamma(x)\,\Gamma(x+\tfrac{1}{3})\,\Gamma(x+\tfrac{2}{3}) = \int_0^\infty\int_0^\infty\int_0^\infty e^{-(t+s+r)}t^{x-1}s^{x-2/3}u^{x-1/3}\mathrm{d}t\,\mathrm{d}s\,\mathrm{d}u\,.$$

Change variable $t \to \tau$ as $t = \frac{1}{s\,u}\,\tau^3$ leads to

$$\Gamma(x)\,\Gamma(x+\tfrac{1}{3})\,\Gamma(x+\tfrac{2}{3}) = 3\int_0^\infty\int_0^\infty\int_0^\infty e^{-(\frac{\tau^3}{s\,u}+s+u)}\,\tau^{3x-1}s^{-2/3}u^{-1/3}\mathrm{d}\tau\,\mathrm{d}s\,\mathrm{d}u\,.$$

Repeating, with the appropriate adjustments, the argument of Theorem 2.15, we separate (Fubini's theorem) the integration by integrating first in the variables s and u and then in τ and, at the same time, we use the technique of differentiation of parametric integrals, to evaluate the double integral in s and u, thus introducing:

$$I_2(\tau) = \iint_{[0,+\infty)^2} e^{-(\frac{\tau^3}{s\,u}+s+u)}\,s^{-2/3}u^{-1/3}\,\mathrm{d}s\,\mathrm{d}u\,,$$

we can rewrite left-hand side of (2.51) as:

$$\Gamma(x)\,\Gamma(x+\tfrac{1}{3})\,\Gamma(x+\tfrac{2}{3}) = 3\int_0^\infty \tau^{3x-1}\,I_2(\tau)\,\mathrm{d}\tau\,. \tag{2.52}$$

Now, differentiating assuming τ as the leading variable, we get

$$I_2'(\tau) = -3\tau^2 \iint_{[0,+\infty)^2} e^{-(\frac{\tau^3}{s\,u}+s+u)}\, s^{-5/3} u^{-4/3}\, \mathrm{d}s\, \mathrm{d}u.$$

Then change variable $s \to \sigma$ putting $s = \dfrac{\tau^3}{u\,\sigma}$ so that

$$I_2'(\tau) = -3 \iint_{[0,+\infty)^2} e^{-(\frac{\tau^3}{\sigma\,u}+\sigma+u)}\, u^{-2/3}\, \sigma^{-1/3}\, \mathrm{d}\sigma\, \mathrm{d}u\,.$$

Since the variable of integration is "dummy", the last equality reads as:

$$I_2'(\tau) = -3\, I_2(\tau)\,.$$

Once again we obtain the value of $I_2(\tau)$ by solving a separable differential equation, namely: $I_2(\tau) I_2(0)\, e^{-3\tau}$ where:

$$\begin{aligned} I_2(0) &= \iint_{[0,+\infty)^2} e^{-(s+u)}\, s^{-2/3}\, u^{-1/3}\, \mathrm{d}s\, \mathrm{d}u = \int_0^{+\infty} e^{-s} s^{-2/3}\, \mathrm{d}s \int_0^{+\infty} e^{-u} u^{-1/3}\, \mathrm{d}u \\ &= \Gamma(\tfrac{1}{3})\,\Gamma(\tfrac{2}{3}) = \frac{\pi}{\sin(\frac{\pi}{3})} = \frac{2\pi}{\sqrt{3}}\,. \end{aligned} \tag{2.53}$$

Notice that (2.53) follows from (2.32). Henceforth $I_2(\tau) = \frac{2\pi}{\sqrt{3}}\, e^{-3\tau}$, thus, going back to (2.52) we see that

$$\Gamma(x)\,\Gamma(x+\tfrac{1}{3})\,\Gamma(x+\tfrac{2}{3}) = 2\pi\sqrt{3} \int_0^{\infty} \tau^{3x-1}\, e^{-3\tau}\, \mathrm{d}\tau = 2\pi\, 3^{1-3x}\, \Gamma(3x)\,,$$

completing our proof. □

After implementing the demonstration methodology in two specific but significant situations, before moving on to the general case to provide the proof of Theorem 2.14, we need a formula, again due to Euler, which extends the reflection formula. First, we need a proper trigonometric identity.

Theorem. For $n \in \mathbb{N}$, the following identity holds:

$$\prod_{k=1}^{n-1} \sin\left(\tfrac{k}{n}\pi\right) = n\, 2^{1-n}\,. \tag{2.54}$$

Proof. Call ξ_k, $k = 0, 1, \ldots n-1$ the solutions of equation $x^n = 1$, i.e.:

$$\xi_k = \cos\left(\tfrac{2k}{n}\pi\right) + i\,\sin\left(\tfrac{2k}{n}\pi\right) = e^{\frac{2k}{n}\pi\, i}\,.$$

Hence:

$$x^n - 1 = (x-1) \sum_{k=1}^{n-1} x^k = \prod_{k=0}^{n-1} (x - \xi_k)\,.$$

Simplifying $x-1$, we obtain, restarting the initial index in the product due to $\xi_0 = 1$

$$\sum_{k=1}^{n-1} x^k = \prod_{k=1}^{n-1} (x - \xi_k)\,. \tag{P}$$

Taking $x = 1$ in (P), we arrive at

$$n = \prod_{k=1}^{n-1} (1 - \xi_k)\,. \tag{U}$$

Write the generic element of the product (2.54) as:

$$\sin\left(\tfrac{k}{n}\pi\right) = (2i)^{-1} e^{-\frac{k}{n}\pi\, i}\left(e^{\frac{2k}{n}\pi\, i} - 1\right) = (2i)^{-1} e^{-\frac{k}{n}\pi\, i}\,(\xi_k - 1)\,.$$

Multiplying

$$\prod_{k=1}^{n-1} \sin\left(\tfrac{k}{n}\pi\right) = (2i)^{1-n} \prod_{k=1}^{n-1} e^{-\frac{k}{n}\pi\, i}\,(\xi_k - 1)\,.$$

On the other side:

$$\begin{aligned}\prod_{k=1}^{n-1} e^{-\frac{k}{n}\pi\, i} &= e^{-\frac{\pi}{n}\, i \sum_{k=1}^{n-1} k} = e^{-\frac{\pi}{n}\, i \frac{n(n-1)}{2}} = e^{-i\,\frac{\pi}{2}(n-1)} \\ &= \left(e^{-i\,\frac{\pi}{2}}\right)^{n-1} = (-i)^{n-1}\,. \end{aligned} \tag{H}$$

In conclusion, from (H) and (U), we infer

$$\prod_{k=1}^{n-1} \sin\left(\tfrac{k}{n}\pi\right) = (2i)^{1-n}(-i)^{n-1} n = n\, 2^{1-n}\,.$$

□

We can demonstrate the following Euler identity on a particular product of gamma values.

Theorem 2.16. For any $n \in \mathbb{N}$

$$\Gamma\left(\tfrac{1}{n}\right)\Gamma\left(\tfrac{2}{n}\right)\cdots\Gamma\left(\tfrac{n-1}{n}\right) = \sqrt{\frac{(2\pi)^{n-1}}{n}}. \tag{2.55}$$

Proof. Thesis follows from Euler's reflection and (2.54), since

$$\begin{aligned}\Gamma\left(\tfrac{1}{n}\right)\Gamma\left(\tfrac{n-1}{n}\right) &= \frac{\pi}{\sin\frac{\pi}{n}}, \\ \Gamma\left(\tfrac{2}{n}\right)\Gamma\left(\tfrac{n-2}{n}\right) &= \frac{\pi}{\sin\frac{2\pi}{n}}, \\ \Gamma\left(\tfrac{3}{n}\right)\Gamma\left(\tfrac{n-3}{n}\right) &= \frac{\pi}{\sin\frac{3\pi}{n}}, \\ \cdots\cdots\cdots &= \cdots \\ \Gamma\left(\tfrac{n-1}{n}\right)\Gamma\left(\tfrac{1}{n}\right) &= \frac{\pi}{\sin\frac{(n-1)\pi}{n}}. \end{aligned}$$

Multiplying again

$$\prod_{k=1}^{n-1} \Gamma^2\left(\tfrac{k}{n}\right) = \frac{\pi^{n-1}}{\displaystyle\prod_{k=1}^{n-1} \sin\left(\tfrac{k\pi}{n}\right)}.$$

The formula (2.55) follows from (2.54). □

Everything is ready for the demonstration of the multiplication formula in the general case, stated in Theorem 2.14. We start from the right-hand side of (2.49), expressing this with an n-dimensional integral, where, for greater readability, this time, we express the integration variables in indexed notation:

$$\Gamma(x)\Gamma(x+\tfrac{1}{n})\dots\Gamma(x+\tfrac{n-1}{n}) =$$
$$\int_0^\infty \cdots \int_0^\infty e^{-(t_1+\cdots+t_n)} t_1^{x-1}\, t_2^{x-\frac{n-1}{n}} \dots t_n^{x-\frac{1}{n}}\, \mathrm{d}t_1 \cdots \mathrm{d}t_n\,.$$

Now change variable $t_1 \to \tau_1$ setting $t_1 = \dfrac{\tau_1^n}{t_2 \dots t_n} \implies \mathrm{d}t_1 = n\frac{\tau^{n-1}}{t_2\cdots t_n}\,\mathrm{d}\tau_1\,.$ Thus, the given n-dimensional integral is changed into

$$n\int_0^\infty \cdots \int_0^\infty e^{-\left(\frac{\tau_1^n}{t_2\dots t_n}+t_2+\cdots+t_n\right)} \tau_1^{nx-1} t_2^{-\frac{n-1}{n}} \cdots t_n^{-\frac{1}{n}}\, \mathrm{d}\tau_1 \mathrm{d}t_2 \cdots \mathrm{d}t_n\,.$$

Exchanging the order of integration, we obtain

$$n\int_0^\infty \tau_1^{nx-1} \left(\int_0^\infty e^{-\left(\frac{\tau_1^n}{t_2\dots t_n}+t_2+\cdots+t_n\right)} t_2^{-\frac{n-1}{n}} \cdots t_n^{-\frac{1}{n}}\, \mathrm{d}t_2 \cdots \mathrm{d}t_n \right) \mathrm{d}\tau_1\,.$$

For simplicity, the integral inside the round bracket must be understood as $n-1$ dimensional: let's compute it thinking of it as dependent on τ_1 acting as a parameter via differentiation. Thus, we set the following:

$$I_{n-1}(\tau_1) = \int_0^\infty e^{-\left(\frac{\tau_1^n}{t_2\dots t_n}+t_2+\cdots+t_n\right)} t_2^{-\frac{n-1}{n}} \cdots t_n^{-\frac{1}{n}}\, \mathrm{d}t_2 \cdots \mathrm{d}t_n\,.$$

Differentiating, we obtain

$$\begin{aligned} I'_{n-1}(\tau_1) &= -n \int_0^\infty \frac{\tau_1^{n-1}}{t_2\dots t_n} e^{-\left(\frac{\tau_1^n}{t_2\dots t_n}+t_2+\cdots+t_n\right)} t_2^{x-\frac{n-1}{n}} \cdots t_n^{x-\frac{1}{n}}\, \mathrm{d}t_2 \cdots \mathrm{d}t_n \\ &= -n \int_0^\infty e^{-\left(\frac{\tau_1^n}{t_2\dots t_n}+t_2+\cdots+t_n\right)} \tau_1^{n-1} t_2^{-\frac{n-1}{n}-1} t_3^{-\frac{n-2}{n}-1} \cdots t_n^{-\frac{1}{n}-1}\, \mathrm{d}t_2 \cdots \mathrm{d}t_n\,. \end{aligned}$$

The change of variable $t_2 \to \tau_2$, $t_2 = \frac{\tau_1^n}{\tau_2} \frac{1}{t_3\cdots t_n}$, leads to

$$I'_{n-1}(\tau_1) = -n \int_0^\infty e^{-\left(\frac{\tau_1^n}{\tau_2\, t_3\dots t_n}+\tau_2+t_3+\cdots+t_n\right)} \tau_2^{-\frac{1}{n}} t_3^{-\frac{n-1}{n}} \dots t_n^{-\frac{2}{n}}\, \mathrm{d}\tau_2 \mathrm{d}t_3 \cdots \mathrm{d}t_n\,.$$

Again, because the integration variables are immaterial, we can conclude that the parametric integral $I_{n-1}(\tau_1)$ satisfies the differential equation.

$$I'_{n-1}(\tau_1) = -nI_{n-1}(\tau_1) \implies I_{n-1}(\tau_1) = I_{n-1}(0)\, e^{-n\tau_1}.$$

But

$$\begin{aligned} I_{n-1}(0) &= \int_0^\infty e^{-(t_2+\cdots+t_n)}\, t_2^{-\frac{n-1}{n}} \cdots t_n^{-\frac{1}{n}}\, \mathrm{d}t_2 \cdots \mathrm{d}t_n \\ &= \int_0^\infty e^{-t_2}\, t_2^{-\frac{n-1}{n}}\, \mathrm{d}t_2 \cdots \int_0^\infty e^{-t_n}\, t_n^{-\frac{1}{n}}\, \mathrm{d}t_n \\ &= \Gamma(\tfrac{1}{n}) \cdots \Gamma(\tfrac{n-1}{n}) = \sqrt{\tfrac{(2\pi)^{n-1}}{n}}. \end{aligned}$$

In the last step, we used Theorem 2.16. We are about to end our proof since:

$$\begin{aligned} \Gamma(x)\Gamma(x+\tfrac{1}{n})\ldots\Gamma(x+\tfrac{n-1}{n}) &= n\int_0^\infty \tau_1^{nx-1} I_{n-1}(\tau_1)\mathrm{d}\tau_1 \\ &= n\int_0^\infty \tau_1^{nx-1}\sqrt{\frac{(2\pi)^{n-1}}{n}}\, e^{-n\tau_1}\,\mathrm{d}\tau_1 \\ &= \sqrt{n}\,\sqrt{(2\pi)^{n-1}}\int_0^\infty \tau_1^{nx-1} e^{-n\tau_1}\,\mathrm{d}\tau_1 \\ &= \sqrt{n}\,\sqrt{(2\pi)^{n-1}}\int_0^\infty e^{-t} n^{-nx} t^{nx-1}\,\mathrm{d}t. \qquad (\dagger) \\ &= n^{\frac{1}{2}-nx}\,(2\pi)^{\frac{n-1}{2}}\,\Gamma(nx) \end{aligned}$$

Notice that in (†) we change variable $\tau_1 \to t$ setting $n\tau_1 = t$, completing the proof of Theorem 2.14. □

2.8 Representing gamma by infinite product

We begin with a representation formula due to Euler and Gauss.

Theorem 2.17. If $x > 0$ then

$$\Gamma(x) = \lim_{n\to\infty} \frac{n!\, n^x}{x(x+1)\cdots(x+n)}. \qquad (2.56)$$

Proof. We begin with the integral:

$$\begin{aligned} \int_0^1 t^{x-1}(1-t)^n \mathrm{d}t = \mathrm{B}(x, n+1) &= \frac{\Gamma(x)\Gamma(n+1)}{\Gamma(x+n+1)} \\ &= \frac{n!\,\Gamma(x)}{(x+n)(x+n-1)\cdots x\,\Gamma(x)} = \frac{n!}{(x+n)(x+n-1)\cdots x} \end{aligned}$$

In the integral on the left-hand side, we change the variable t/n instead of t, obtaining:

$$\frac{n!}{(x+n)(x+n-1)\cdots x} = \frac{1}{n^x}\int_0^n t^{x-1}\left(1-\frac{t}{n}\right)^n \mathrm{d}t\,.$$

So, by introducing the sequence of functions:

$$g_n(t) = \begin{cases} \left(1-\dfrac{t}{n}\right)^n & \text{for } 0 \le t \le n, \\ 0 & \text{for } n < t < \infty, \end{cases}$$

we have

$$\frac{n!\,n^x}{(x+n)(x+n-1)\cdots x} = \int_0^\infty t^{x-1} g_n(t)\mathrm{d}t.$$

Notice that the (increasing) sequence of functions $g_n(t)$ converges as $n \to \infty$ to e^{-t}. The exchange between the limit and the integral is legitimate for Dini's Theorem 1.3. Therefore, we infer

$$\lim_{n\to\infty} \frac{n!\,n^x}{(x+n)(x+n-1)\cdots x} = \int_0^\infty t^{x-1}e^{-t}\mathrm{d}t = \Gamma(x).$$

The proof of (2.56) is completed. □

The formula (2.56) can be rewritten in terms of an infinite product. Observing that

$$\prod_{k=1}^{n-1}\left(1+\frac{1}{k}\right) = n$$

we infer

$$\prod_{k=1}^{n-1}\left(1+\frac{1}{k}\right)^x = n^x.$$

Hence, we obtain

$$\frac{n!n^x}{x(x+1)\cdots(x+n)} = \frac{1}{x}\left(1+\frac{1}{n}\right)^{-x}\prod_{k=1}^{n}\frac{\left(1+\frac{1}{k}\right)^x}{1+\frac{x}{k}}.$$

Thus taking the limit for $n \to +\infty$ and recalling (2.56) we arrive at

$$\frac{1}{x}\prod_{k=1}^{\infty}\frac{\left(1+\frac{1}{k}\right)^x}{1+\frac{x}{k}} = \Gamma(x),$$

In other words, we obtained, recalling the functional property (2.4)

$$\prod_{n=1}^{\infty}\frac{\left(1+\frac{1}{n}\right)^x}{1+\frac{x}{n}} = \Gamma(1+x). \tag{2.57}$$

The infinite product appearing on the left-hand side of (2.57) converges uniformly on every compact of the complex plane that does not contain negative integers. This infinite product is, therefore, the analytical continuation, see definition 1.16 page 17, of the Legendre integral (2.2). It also demonstrates that the infinite product appearing on the right-hand side of (2.57) converges uniformly on every compact of the complex plane that does not contain negative integers.

It is worth noting that the analytical continuation of gamma provided by (2.56) and (2.57) proves that Γ has no zeros in the complex plane. It is possible to evaluate the residues at the negative integers that are simple poles of gamma.

Theorem.

$$\operatorname{Res}(\Gamma, -n) = \frac{(-1)^n}{n!} \tag{2.58}$$

Proof. To obtain (2.58) we need to evaluate

$$\ell = \lim_{x \to -n} (x+n)\,\Gamma(x).$$

Observing that

$$\begin{aligned}(x+n)\,\Gamma(x) &= (x+n)\left(\frac{x(x+1)\dots(x+n)}{x(x+1)\dots(x+n)}\right)\Gamma(x),\\ &= (x+n)\left(\frac{\Gamma(x+n+1)}{x(x+1)\dots(x+n)}\right), &\text{(a)}\\ &= \left(\frac{\Gamma(x+n+1)}{x(x+1)\dots(x+n-1)}\right). &\text{(b)}\end{aligned}$$

Where (a) follows from (2.12) and (b) follows by a straightforward cancellation. The limit evaluation is elementary and leads to (2.58). □

Remark. From the formula (2.56) and the reflection formula (2.32), provide an alternative demonstration of the infinite product for the sine. In fact, choosing $0 < x < 1$ and observing that

$$\Gamma(1-x) = \lim_{n\to\infty} \frac{n!n^x}{(1-x)(2-x)\cdots(n-x)(n+1-x)} \tag{2.59}$$

moreover, by multiplying the terms of the two sequences on the right-hand side of (2.56) and (2.59), we get

$$\frac{(n!)^2\, n}{x(1+x)(1-x)(2+x)(2-x)\dots(n+x)(n-x)(n+1-x)} =$$

$$\frac{(n!)^2\, n}{x(1-x^2)(2^2-x^2)\dots(n^2-x^2)(n+1-x)} =$$

$$\frac{n}{x(1-\frac{x^2}{1^2})(1-\frac{x^2}{2^2})\dots(1-\frac{x^2}{n^2})(n+1-x)} =$$

$$\frac{1}{x\displaystyle\prod_{k=1}^{n}\left(1-\frac{x^2}{k^2}\right)(1+\dfrac{1-x}{n})}.$$

Therefore

$$\Gamma(x)\Gamma(1-x) = \lim_{n\to\infty} \frac{1}{x\prod_{k=1}^{n}\left(1-\frac{x^2}{k^2}\right)(1+\dfrac{1-x}{n})}$$
$$= \frac{1}{x\prod_{n=1}^{\infty}\left(1-\frac{x^2}{n^2}\right)}. \tag{2.60}$$

Comparing (2.60) with the reflection formula (2.32) leads to our statement, in fact we have

$$\prod_{n=1}^{\infty}\left(1-\frac{x^2}{n^2}\right) = \frac{\sin(\pi\, x)}{\pi\, x}.$$

which agrees with (1.41). □

To demonstrate Weierstrass's representation Theorem 2.19, we need to introduce the Euler–Mascheroni constant[1] which is defined by the limit:

$$\gamma = \lim_{n\to\infty}\left(\sum_{k=1}^{n}\frac{1}{k} - \ln n\right). \tag{2.61}$$

The existence of the limit (2.61) is a consequence of the following theorem due to Euler and Maclaurin.

Theorem 2.18. If $f : [0,\infty[\to\mathbb{R}$ is positive, continuous and decreasing, then defining for each $n\in\mathbb{N}$

$$\varphi_n := \sum_{k=1}^{n} f(k) - \int_1^n f(x)\mathrm{d}x\,, \tag{2.62}$$

we have that (φ_n) converges as $n\to+\infty$.

Proof. The thesis is proved by demonstrating that (φ_n) is decreasing, non-negative and bounded from below. Recall that if $f : [a,b]\to\mathbb{R}$ is continuous, then

$$m(b-a) \le \int_a^b f(x)\mathrm{d}x \le M(b-a), \tag{2.63}$$

where

$$m = \inf_{x\in[a,b]} f(x) = \min_{x\in[a,b]} f(x), \quad M = \sup_{x\in[a,b]} f(x) = \max_{x\in[a,b]} f(x)\,.$$

[1]Lorenzo Mascheroni (1750–1800) in 1790, published "Adnotationes ad calculum Integrale Euleri" in which he calculated the first 32 digits of the decimal development of the constant, while the first 16 digits were calculated by Euler in 1781.

Thus taking $x \in [k, k+1]$, recalling that $f(x)$ is decreasing, from (2.63) follows that

$$f(k+1) \le \int_k^{k+1} f(x)\mathrm{d}x \le f(k)\,. \tag{2.63b}$$

Moreover, from (2.63b) we infer

$$\begin{aligned}\varphi_{n+1} - \varphi_n &= \left(\sum_{k=1}^{n+1} f(k) - \int_1^{n+1} f(x)\mathrm{d}x\right) - \left(\sum_{k=1}^{n} f(k) - \int_1^{n} f(x)\mathrm{d}x\right) \\ &= f(n+1) - \int_n^{n+1} f(x)\mathrm{d}x \le f(n+1) - f(n+1) = 0\,.\end{aligned}$$

So (φ_n) is decreasing. Finally, (φ_n) is bounded from below, since

$$\begin{aligned}\varphi_n &= \sum_{k=1}^{n} f(k) - \int_1^{n} f(x)\mathrm{d}x = \sum_{k=1}^{n} f(k) - \sum_{k=1}^{n-1}\int_k^{k+1} f(x)\mathrm{d}x \\ &\ge \sum_{k=1}^{n} f(k) - \sum_{k=1}^{n-1} f(k) = f(n) \ge 0.\end{aligned}$$

□

Remark. The existence of the limit (2.61) a special case of the Theorem 2.18 in the case $f(x) = 1/x$.

After these premises, we can demonstrate the Weierstrass infinite representation for gamma.

Theorem 2.19. If γ is given be (2.62), then for any $x \in \mathbb{C}$

$$\frac{1}{\Gamma(x)} = x e^{\gamma x} \prod_{n=1}^{\infty} \left(1 + \frac{x}{n}\right) e^{-\frac{x}{n}}\,. \tag{2.64}$$

Proof. Firstly, we observe that the infinite product on the left-hand side in (2.64) converges for any $x \in \mathbb{C}$ since, expanding the exponential, we find

$$\begin{aligned}\left(1 + \frac{x}{n}\right) e^{-\frac{x}{n}} &= \left(1 + \frac{x}{n}\right)\left(1 - \frac{x}{n} + \frac{x^2}{2n^2} + \cdots\right) \\ &= 1 - \frac{x^2}{2n^2} + \mathrm{O}\left(\frac{1}{n^3}\right)\end{aligned}$$

and series $\sum 1/n^2$ converges. From formula (2.56) proved in Theorem 2.17, we can write:

$$\begin{aligned}\frac{1}{\Gamma(x)} &= \lim_{n\to\infty} \frac{x(x+1)\cdots(x+n)}{n!\,n^x}\\ &= \lim_{n\to\infty} x\left(1+\frac{x}{1}\right)\left(1+\frac{x}{2}\right)\cdots\left(1+\frac{x}{n}\right)n^{-x}\\ &= \lim_{n\to\infty} xe^{-x\ln n}\prod_{k=1}^{n}\left(1+\frac{x}{k}\right)\\ &= \lim_{n\to\infty} xe^{x\left(1+\frac{1}{2}+\cdots+\frac{1}{n}-\ln n\right)}\prod_{k=1}^{n}\left(1+\frac{x}{k}\right)e^{-\frac{x}{k}}\\ &= xe^{\gamma x}\prod_{n=1}^{\infty}\left(1+\frac{x}{n}\right)e^{-\frac{x}{n}}\end{aligned}$$

□

The representation by infinite product of the reciprocal of gamma (2.64) leads to the following graphical representation, which highlights the zeros of $1/\Gamma$ and, therefore, the poles of gamma.

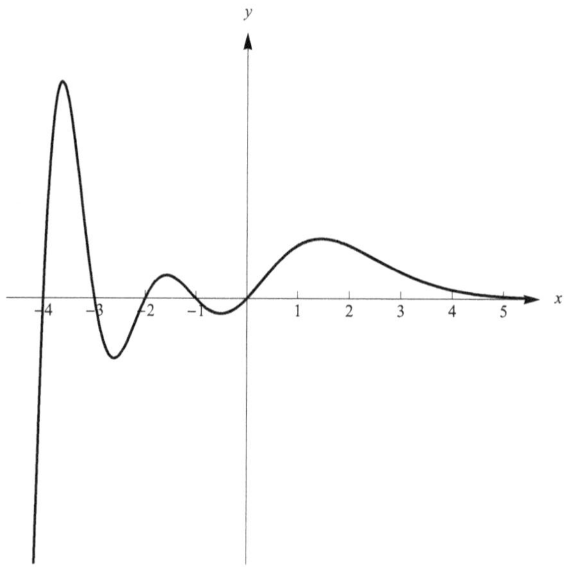

Figure 2.2: Plot of $1/\Gamma(x)$

Corollary 2.20. If $x \in \mathbb{C}$, $x \neq 0, -1, -2, \ldots$ then

$$\Gamma(x) = \frac{e^{-\gamma x}}{x}\prod_{n=1}^{\infty}\left(1+\frac{x}{n}\right)^{-1} e^{\frac{x}{n}}. \tag{2.65}$$

As observed in [4], formulas (2.64) allow, given Theorem 2.9, the analytical continuation (see definition 1.16 of Beta, again through its infinite product representation.

Corollary. If x, y are not zero or negative integers, then

$$B(x,y) = \frac{x+y}{xy} \prod_{n=1}^{\infty} \frac{\left(1+\frac{x+y}{n}\right)}{\left(1+\frac{x}{n}\right)\left(1+\frac{y}{n}\right)} \tag{2.66}$$

Proof. The formula (2.66) follows by substituting into (2.28) formula (2.65). □

2.9 Bohr-Mollerup-Artin theorem

This section is devoted to the fundamental fact that the gamma function is uniquely determined by the functional equation $\Gamma(x+1) = x\Gamma(x)$ together with the initial condition $\Gamma(1) = 1$ and the convexity of $\ln \Gamma(x)$, see Theorem 2.1. Harald Bohr[2] (1887–1951) and Johannes Mollerup (1872–1937) [22] found that these properties uniquely determine the gamma function. Bohr and Mollerup's reasoning was later clarified and brought to the attention of the scientific community by Emil Artin (1898–1962).

Theorem. Assume $G(x)$ is a positive function with domain $x > 0$ such that

i) $G(x+1) = xG(x)$ ii) $G(1) = 1$ iii) $\ln G(x)$ convex

then $G(x) = \Gamma(x)$.

Proof. The proof presented here is due to [39, pages 258–259]. From hypothesis i), it follows that for $n \in \mathbb{N}$, $G(n+1) = n!$. Observed that for each $n \in \mathbb{N}$ and for $0 < x \leq 1$ the following identity holds:

$$n + x = (1-x)\,n + x\,(n+1). \tag{2.67}$$

Notice that (2.67) expresses the number $n+x$ as a convex combination of the successive integers n and $n+1$, hence the convexity assumption iii) yields:

$$\ln G(n+x) \leq (1-x)\ln G(n) + x \ln G(n+1)\,. \tag{2.68}$$

Formula (2.68), using the properties of the logarithm, is written as

$$\begin{aligned} G(n+x) &\leq G^{1-x}(n)G^{x}(n+1) = [(n-1)!]^{1-x}[n!]^{x} \\ &= \left[\frac{n!}{n}\right]^{1-x}[n!]^{x} = n!\,n^{x-1}\,. \end{aligned} \tag{2.69}$$

[2]Harald Bohr is the younger brother of Niels, Nobel laureate in Physics in 1922.

Similarly, the convex combination.

$$n+1 = x(n+x) + (1-x)(n+x+1)$$

implies inequality:

$$n! = G(n+1) \leq G^x(n+x)G^{1-x}(n+x+1) = G(n+x)(n+x)^{1-x}, \quad (2.70)$$

since $G(n+x+1) = (n+x)G(n+x)$ by virtue of property (i). Combining the inequalities (2.69) and (2.70) we find

$$n!(n+x)^{x-1} \leq G(n+x) \leq n!n^{x-1}. \quad (2.71)$$

On the other hand, property (i) can be iterated by providing the following:

$$G(x+n) = x(x+1)\cdots(x+n-1)G(x)$$

and thus the inequality (2.71) takes the form:

$$\frac{n!(n+x)^{x-1}}{x(x+1)\cdots(x+n-1)} \leq G(x) \leq \frac{n!n^{x-1}}{x(x+1)\cdots(x+n-1)}. \quad (2.72)$$

Now as $n \to \infty$ we can invoke the Theorem 2.17, equation (2.56) in (2.72) to conclude that

$$\Gamma(x) \leq G(x) \leq \Gamma(x)$$

so that $G(x) = \Gamma(x)$ for $0 < x \leq 1$. The extension to all $x > 0$ follows from the functional identities $G(x+1) = xG(x)$ and $\Gamma(x+1) = x\Gamma(x)$. □

Remark. The inequality (2.72) can be used to prove the existence of the limit (2.56) in Theorem 2.17 and to show that it is equal to $G(x)$. Moreover, since the reasoning is also valid for $\Gamma(x)$ in addition to the resulting equality $G(x) = \Gamma(x)$ we have an alternative proof of Theorem 2.17.

2.10 Definite integrals in terms of Eulerian integrals

Eulerian functions are of enormous use in the computation of remarkable definite integrals and the summation of infinite series, leading to beneficial applications in studying various random variables. The first integral identity is due to Legendre [75].

Theorem. If $n \geq 1$ then:

$$\int_0^1 \frac{dx}{\sqrt{1-x^n}} = \cos\left(\frac{\pi}{n}\right) \int_0^1 \frac{dx}{\sqrt{1+x^n}}. \quad (2.73)$$

Proof. First, observe that, for $n \geq 1$, both integrals converge. Define:

$$I_1 = \int_0^1 \frac{\mathrm{d}x}{\sqrt{1-x^n}}, \quad I_2 = \int_0^1 \frac{\mathrm{d}x}{\sqrt{1+x^n}}.$$

Change variable: $x^n = t$, in both integrals, so that

$$I_1 = \frac{1}{n}\int_0^1 \frac{t^{\frac{1}{n}-1}}{\sqrt{1-t}}\,\mathrm{d}t = \frac{1}{n}\,\mathrm{B}\left(\frac{1}{n},\frac{1}{2}\right)$$

$$I_2 = \frac{1}{n}\int_0^\infty \frac{t^{\frac{1}{n}-1}}{\sqrt{1+t}}\,\mathrm{d}t = \frac{1}{n}\,\mathrm{B}\left(\frac{1}{n},\frac{1}{2}-\frac{1}{n}\right).$$

In I_2, above, we used the Beta representation (2.25b). Now, form the ratio I_1/I_2 and recall Theorem 2.9, to get

$$\frac{I_1}{I_2} = \frac{\mathrm{B}\left(\frac{1}{n},\frac{1}{2}\right)}{\mathrm{B}\left(\frac{1}{n},\frac{1}{2}-\frac{1}{n}\right)} = \frac{\Gamma\left(\frac{1}{n}\right)\Gamma\left(\frac{1}{2}\right)}{\Gamma\left(\frac{1}{n}+\frac{1}{2}\right)} \quad \frac{\Gamma\left(\frac{1}{2}\right)}{\Gamma\left(\frac{1}{n}\right)\Gamma\left(\frac{1}{2}-\frac{1}{n}\right)}$$

$$= \frac{\pi}{\Gamma\left(\frac{1}{n}+\frac{1}{2}\right)\Gamma\left(\frac{1}{2}-\frac{1}{n}\right)}.$$

Thus, recalling (2.34):

$$\frac{I_1}{I_2} = \frac{\pi}{\dfrac{\pi}{\cos\left(\dfrac{\pi}{n}\right)}} = \cos\left(\frac{\pi}{n}\right),$$

hence formula (2.73) is proved. □

This technique can be applied to the next identity (2.74), which we leave to the reader.

Theorem. If $2a < n$, then:

$$\int_0^1 \frac{x^{a-1}}{\sqrt{1-x^n}}\,\mathrm{d}x = \cos\left(\frac{a}{n}\,\pi\right)\int_0^\infty \frac{z^{a-1}}{\sqrt{1+z^n}}\,\mathrm{d}z. \tag{2.74}$$

Another interesting integrals are the following (2.75), (2.76), and (2.77).

Theorem. If $n \in \mathbb{N}$, $n \geq 2$, then

$$\int_0^1 \frac{\mathrm{d}x}{\sqrt[n]{1-x^n}} = \frac{\pi}{n\,\sin\left(\dfrac{\pi}{n}\right)}. \tag{2.75}$$

Proof. Using, again, the change of variable $x^n = t$, we obtain

$$\int_0^1 \frac{\mathrm{d}x}{\sqrt[n]{1-x^n}} = \frac{1}{n}\int_0^1 t^{\frac{1}{n}-1}(1-t)^{-\frac{1}{n}}\,\mathrm{d}t = \frac{1}{n}\int_0^1 t^{\frac{1}{n}-1}(1-t)^{\left(1-\frac{1}{n}\right)-1}\,\mathrm{d}t$$
$$= \frac{1}{n}\mathrm{B}\left(\frac{1}{n}, 1-\frac{1}{n}\right) = \frac{1}{n}\Gamma\left(\frac{1}{n}\right)\Gamma\left(1-\frac{1}{n}\right).$$

Thesis (2.75) follows from the reflection formula (2.32). □

Theorem. For any $n \geq 2$ the following identity holds:

$$\int_0^\infty \frac{\mathrm{d}x}{1+x^n} = \frac{\pi}{n\,\sin\left(\dfrac{\pi}{n}\right)}. \tag{2.76}$$

Moreover, if $n - m > 1$, then

$$\int_0^\infty \frac{x^m}{1+x^n}\,\mathrm{d}x = \frac{\pi}{n\,\sin\left(\dfrac{\pi}{n}(m+1)\right)}. \tag{2.77}$$

Proof. The change of variable $1 + x^n = t^{-1}$ is employed on the left-hand-side integral of formula (2.76), and therefore $\mathrm{d}x = -1/n\,t^{-\frac{1}{n}-1}\,(1-t)^{\frac{1}{n}-1}\,\mathrm{d}t$:

$$\int_0^\infty \frac{\mathrm{d}x}{1+x^n} = \frac{1}{n}\int_0^1 t^{-\frac{1}{n}}(1-t)^{\frac{1}{n}-1}\,\mathrm{d}t = \frac{1}{n}\int_0^1 t^{\left(1-\frac{1}{n}\right)-1}(1-t)^{\frac{1}{n}-1}\,\mathrm{d}t$$
$$= \frac{1}{n}\mathrm{B}\left(1-\frac{1}{n}, \frac{1}{n}\right) = \frac{1}{n}\Gamma\left(1-\frac{1}{n}\right)\Gamma\left(\frac{1}{n}\right).$$

(2.76) follows from (2.32). An analogous argument gives (2.77). □

To prove the following formula (2.78), we need formula (2.24), proved in Theorem 2.8.

Theorem. If $n \in \mathbb{N}$, then:

$$\int_{-\infty}^\infty \frac{\mathrm{d}x}{(1+x^2)^n} = \frac{\pi}{2^{n-1}}\frac{(2n-3)!!}{(n-1)!}. \tag{2.78}$$

Proof. Due to the symmetry of the integrand, we have

$$\int_{-\infty}^\infty \frac{\mathrm{d}x}{(1+x^2)^n} = 2\int_0^\infty \frac{\mathrm{d}x}{(1+x^2)^n}.$$

The change of variable $1 + x^2 = t^{-1}$ leads to

$$\int_{-\infty}^\infty \frac{\mathrm{d}x}{(1+x^2)^n} = \mathrm{B}\left(n-\frac{1}{2}, \frac{1}{2}\right) = \frac{\sqrt{\pi}}{(n-1)!}\Gamma\left(n-\frac{1}{2}\right).$$

Exploiting (2.24), we arrive at the thesis (2.78). □

By an analogous argument, the following identity (2.79) can be demonstrated.

Theorem. If $np - m > 1$, then

$$\int_0^\infty \frac{x^m}{(1+x^n)^p}\,\mathrm{d}x = \frac{\Gamma\left(\frac{m+1}{n}\right)\Gamma\left(p-\frac{m+1}{n}\right)}{n\,\Gamma(p)}. \tag{2.79}$$

We prove a famous summation formula due to Johann Bernoulli

Theorem.

$$\int_0^1 x^{-x}\mathrm{d}x = \sum_{m=1}^\infty \frac{1}{m^m}. \tag{2.80}$$

Proof. In the left-hand side integral in (2.80), we use the substitution $x = e^{-t}$ so that

$$\int_0^1 x^{-x}\mathrm{d}x = \int_0^\infty (e^{-t})^{-e^{-t}} e^{-t}\mathrm{d}t = \int_0^\infty e^{te^{-t}} e^{-t}\mathrm{d}t$$

Next, we expand in power series the first factor of the last integral, obtaining:

$$e^{te^{-t}} = \sum_{n=0}^\infty \frac{(te^{-t})^n}{n!} = \sum_{n=0}^\infty \frac{t^n e^{-nt}}{n!};$$

so that

$$\int_0^1 x^{-x}\mathrm{d}x = \sum_{n=0}^\infty \frac{1}{n!}\int_0^\infty t^n e^{-(n+1)t}\mathrm{d}t = \sum_{m=1}^\infty \frac{1}{(m-1)!}\int_0^\infty t^{m-1}e^{-mt}\mathrm{d}t$$

Now let's set $mt = s$ so that

$$\int_0^1 x^{-x}\mathrm{d}x = \sum_{m=1}^\infty \frac{1}{(m-1)!}\frac{1}{m^m}\int_0^\infty e^{-s}s^{m-1}\mathrm{d}s = \sum_{m=1}^\infty \frac{1}{(m-1)!}\frac{1}{m^m}\Gamma(m),$$

which shows (2.80). □

2.11 Fresnel and Dirichlet integrals

Double integration often leads to some interesting identities exchanging the integration order. Here, we obtain some remarkable identities, connecting the Eulerian integrals with double integration. We obtain the (famous) Fresnel integrals, see formula (2.89), which are related to the probability integrals, and the Dirichlet integrals see formulas (2.81) and (2.84), following the presentation by [85]. We begin proving the (beautiful) identity (2.81), which for a particular value of p is of great importance in the theory of the zeta function in chapter 4, Section 4.4.

Theorem. For any $b \in \mathbb{R}$ and for $0 < p < 2$, we have

$$\int_0^{+\infty} \frac{\sin(b\,x)}{x^p}\,\mathrm{d}x = \frac{\pi\,b^{p-1}}{2\,\Gamma(p)\,\sin\left(p\,\dfrac{\pi}{2}\right)}. \tag{2.81}$$

Proof. The starting point to prove (2.81), is the double integral:

$$I(b,p) = \int_0^{+\infty}\int_0^{+\infty} \sin(b\,x)\,y^{p-1}\,e^{-x\,y}\,\mathrm{d}x\,\mathrm{d}y.$$

The assumption we made of the parameters ensures summability. Hence, the integration can be performed in any order. Let us integrate, first, beginning from x:

$$I(b,p) = \int_0^{+\infty} y^{p-1}\left(\int_0^{+\infty} \sin(b\,x)\;e^{-x\,y}\,\mathrm{d}x\right)\,\mathrm{d}y\,.$$

The inner integral is elementary, in fact

$$\int_0^{+\infty} \sin(b\,x)\,e^{-x\,y}\,\mathrm{d}x = \frac{b}{b^2+y^2}.$$

Then

$$I(b\,,p) = b\int_0^{+\infty} \frac{y^{p-1}}{b^2+y^2}\,\mathrm{d}y\,.$$

Thus, changing variable putting $t = y/b$, we obtain

$$I(b,p) = b^{p-1}\int_0^{\infty} \frac{t^{p-1}}{1+t^2}\,\mathrm{d}t\,.$$

Hence, we can use (2.77) and complete the first computation:

$$I(b,p) = \frac{\pi\,b^{p-1}}{2\,\sin\left(p\,\frac{\pi}{2}\right)}. \tag{2.82}$$

Now, we reverse the order of integration:

$$I(b,p) = \int_0^{+\infty} \sin(b\,x)\left(\int_0^{+\infty} y^{p-1}\,e^{-x\,y}\,\mathrm{d}y\right)\,\mathrm{d}x.$$

The inner integral is evaluated, in terms of gamma, see proposition 2.2:

$$\int_0^{+\infty} y^{p-1}\,e^{-x\,y}\,\mathrm{d}y = \frac{1}{x^p}\,\Gamma(p). \tag{2.83}$$

Equating (2.83) and (2.82) leads to

$$\Gamma(p)\int_0^{\infty} \frac{\sin(b\,x)}{x^p}\,\mathrm{d}x = \frac{b^{p-1}\pi}{2\,\sin\left(p\,\frac{\pi}{2}\right)},$$

which is nothing else but (2.81). This integral is also evaluated in [39, p. 261]. □

In consequence of equation (2.81), considering the particular case $b = p = 1$, we obtain the Dirichlet integral. Moreover, from (2.81), it is possible to establish a second integral formula; see (2.84) below, which generalises the Dirichlet integral.

Theorem. For $q > 1$, the following formula holds:

$$\int_0^\infty \frac{\sin x^q}{x^q}\, \mathrm{d}x = \frac{1}{q-1}\, \Gamma\left(\frac{1}{q}\right) \cos\left(\frac{\pi}{2\,q}\right). \tag{2.84}$$

Proof. The first step is the change of variable $x^q = u$, in the integral:

$$\int_0^\infty \frac{\sin x^q}{x^q}\, \mathrm{d}x = \frac{1}{q} \int_0^\infty \frac{\sin u}{u^{2-\frac{1}{q}}}\, \mathrm{d}u.$$

The right-hand side integral has the form (2.81), with $b = 1$ and $p = 2 - 1/q$, therefore

$$\int_0^\infty \frac{\sin x^q}{x^q}\, \mathrm{d}x = \frac{\pi}{2\,q\,\Gamma\left(2 - \frac{1}{q}\right) \sin\left(\left(2 - \frac{1}{q}\right)\frac{\pi}{2}\right)}.$$

But, since

$$\sin\left(\left(2 - \frac{1}{q}\right)\frac{\pi}{2}\right) = \sin\left(\pi - \frac{\pi}{2\,q}\right) = \sin\left(\frac{\pi}{2\,q}\right),$$

using the reflection formula, see page 72, we obtain

$$\begin{aligned}\Gamma\left(2 - \frac{1}{q}\right) &= \Gamma\left(1 + 1 - \frac{1}{q}\right) = \left(1 - \frac{1}{q}\right) \Gamma\left(1 - \frac{1}{q}\right)\\ &= \frac{q-1}{q}\, \frac{\pi}{\sin\left(\frac{\pi}{q}\right)}\, \frac{1}{\Gamma\left(\frac{1}{q}\right)}.\end{aligned}$$

In conclusion, we arrive at the following:

$$\int_0^\infty \frac{\sin x^q}{x^q}\, \mathrm{d}x = \frac{1}{2\,(q-1)}\, \Gamma\left(\frac{1}{q}\right) \frac{\sin\left(\frac{\pi}{q}\right)}{\sin\left(\frac{\pi}{2\,q}\right)}.$$

Finally, the identity

$$\frac{\sin x}{\sin \frac{x}{2}} = 2 \cos \frac{x}{2},$$

implies the equality below, which simplifies to (2.84):

$$\int_0^\infty \frac{\sin x^q}{x^q}\, \mathrm{d}x = \frac{1}{2\,(q-1)}\, \Gamma\left(\frac{1}{q}\right) 2 \cos\left(\frac{\pi}{2\,q}\right).$$

□

We now show, employing reversal integration again, a cosine relation similar to (2.81).

Theorem. For $0 < p < 1$ and $b > 0$ we have

$$\int_0^\infty \frac{\cos(b\,x)}{x^p}\,\mathrm{d}x = \frac{\pi\, b^{p-1}}{2\,\Gamma(p)\,\cos\left(p\,\dfrac{\pi}{2}\right)}. \tag{2.85}$$

Proof. To prove (2.85), we consider the double integral:

$$\int_0^\infty \int_0^\infty \cos(b\,x)\,y^{p-1} e^{-x\,y}\,\mathrm{d}x\,\mathrm{d}y,$$

from which we show that (2.85) can be reached by exchanging the order of integration. The starting point is, then, equality

$$\int_0^\infty \cos(b\,x)\left(\int_0^\infty y^{p-1}\,e^{-x\,y}\,\mathrm{d}y\right)\mathrm{d}x = \int_0^\infty y^{p-1}\left(\int_0^\infty \cos(b\,x)\,e^{-x\,y}\,\mathrm{d}x\right)\mathrm{d}y\,. \tag{2.86}$$

The inner integral on the right-hand side of (2.86) is elementary:

$$\int_0^\infty \cos(b\,x)\,e^{-x\,y}\,\mathrm{d}x = \frac{y}{b^2+y^2}.$$

Thus, the right–hand side of (2.86) becomes

$$\int_0^\infty y^{p-1}\,\frac{y}{b^2+y^2}\,\mathrm{d}y = \frac{1}{b^2}\int_0^\infty \frac{y^p}{1+\left(\frac{y}{b}\right)^2}\,\mathrm{d}y = b^{p-1}\int_0^\infty \frac{t^p}{1+t^2}\,\mathrm{d}t.$$

The latter integral, above, is in the form (2.77). Therefore, the right–hand side integral of (2.86) turns out to be, after the change of variable $y = b\sqrt{t}$, using (2.26) and (2.32):

$$\int_0^\infty \frac{y^p}{b^2+y^2}\,\mathrm{d}y = \frac{\pi\,b^{p-1}}{2\,\sin\left(\dfrac{p+1}{2}\,\pi\right)} = \frac{\pi\,b^{p-1}}{2\,\cos\left(\dfrac{p}{2}\,\pi\right)}. \tag{2.87}$$

The inner integral on the left-hand side of (2.86) is given by (2.83). Hence, the left-hand side integral of (2.86) is

$$\Gamma(p)\int_0^\infty \frac{\cos(b\,x)}{x^p}\,\mathrm{d}x. \tag{2.88}$$

Equating (2.87) and (2.88) leads to (2.85). □

There is a further remarkable consequence of equation (2.81), leading to the evaluation of the so-called Fresnel integrals: the first step is given by formula (2.89) below:

Theorem. Assume $b > 0$ and $k > 1$, then

$$\int_0^\infty \sin(b\,x^k)\;dx = \frac{1}{k\;b^{\frac{1}{k}}}\;\Gamma\left(\frac{1}{k}\right)\;\sin\left(\frac{\pi}{2\,k}\right). \tag{2.89}$$

Proof. To prove (2.89), we start by considering its left-hand side integral, using the change of variable $u = x^k$ obtaining:

$$\int_0^\infty \sin(b\,x^k)\,dx = \int_0^\infty \sin(b\,u)\,\frac{1}{k}\,\frac{u^{\frac{1}{k}}}{u}\,du = \frac{1}{k}\int_0^\infty \frac{\sin(b\,u)}{u^{1-\frac{1}{k}}}\,du.$$

The latter integral, above, is in the form (2.81), with $p = 1 - 1/k$. Hence

$$\int_0^\infty \frac{\sin(b\,u)}{u^{1-\frac{1}{k}}}\,du = \frac{\pi\;b^{-\frac{1}{k}}}{2\,\Gamma\left(1-\frac{1}{k}\right)\;\sin\left(\left(1-\frac{1}{k}\right)\frac{\pi}{2}\right)},$$

Thus

$$\int_0^\infty \sin(b\,x^k)\;dx = \frac{\pi}{2\;k\;b^{\frac{1}{k}}\;\Gamma\left(1-\frac{1}{k}\right)\;\sin\left(\frac{\pi}{2}-\frac{\pi}{2\,k}\right)}.$$

We arrive at (2.89), using the reflection formula, page 72, here in the form:

$$\Gamma\left(\frac{1}{k}\right)\;\Gamma\left(1-\frac{1}{k}\right) = \frac{\pi}{\sin\dfrac{\pi}{k}},$$

and the identities:

$$\sin\left(\frac{\pi}{2}-\frac{\pi}{2\,k}\right) = \cos\left(\frac{\pi}{2\,k}\right), \qquad \frac{\sin x}{\cos\dfrac{x}{2}} = 2\;\sin\frac{x}{2}.$$

□

In (2.89), the particular choices $b = 1$ and $k = 2$ correspond to the sine Fresnel integral:

Corollary.

$$\int_0^\infty \sin(x^2)\,dx = \frac{1}{2}\,\Gamma\left(\frac{1}{2}\right)\;\sin\left(\frac{\pi}{4}\right) = \sqrt{\frac{\pi}{8}}.$$

Exploiting the same technique that produced (2.89) from (2.81), it is possible to derive the analogue of (2.90) for the cosine Fresnel integrals.

Theorem. For $b > 0$ and $k > 1$ we have

$$\int_0^\infty \cos(b\,x^k)\;dx = \frac{1}{k\,b^{\frac{1}{k}}}\;\Gamma\left(\frac{1}{k}\right)\;\cos\left(\frac{\pi}{2\,k}\right). \tag{2.90}$$

Proof. To prove (2.90), the starting point is (2.85). Then, as in the sine case, we introduce the change of variable $u = x^k$, we choose $p = 1 - 1/k$, and, via calculations similar to those performed in the sine case, we arrive at

$$\int_0^\infty \cos(b\,x^k)\,\mathrm{d}x = \frac{\pi}{2\,k\,b^{\frac{1}{k}}\,\Gamma\left(1-\dfrac{1}{k}\right)\,\cos\left(\dfrac{\pi}{2}-\dfrac{\pi}{2\,k}\right)}.$$

Formula (2.90) now follows from reflection formula (2.32) here in the form:

$$\Gamma\left(1-\frac{1}{k}\right) = \frac{\pi}{\Gamma\left(\dfrac{1}{k}\right)\,\sin\dfrac{\pi}{k}},$$

and the trigonometric relations:

$$\cos\left(\frac{\pi}{2}-\frac{\pi}{2\,k}\right) = \sin\left(\frac{\pi}{2\,k}\right), \qquad \frac{\sin x}{\sin\dfrac{x}{2}} = 2\,\cos\frac{x}{2}.$$

□

2.12 Laplace and Cauchy integrals

We conclude the chapter by dealing with integrals studied by Laplace and Cauchy. We follow the presentation of Nielsen [87], Sections 63 and 64.

The Laplace integrals were presented by Laplace in his treatise [73, page 471]. Their complex form is given in the following theorem.

Theorem. If $b > 0$ and $0 < x < 1$ and a is any real, then

$$\int_{-\infty}^{\infty} \frac{e^{i\,b\,t}}{(a+i\,t)^x}\,\mathrm{d}t = \frac{2\,\pi\,b^{x-1}}{e^{a\,b}\,\Gamma(x)} \tag{2.91a}$$

$$\int_{-\infty}^{\infty} \frac{e^{-i\,b\,t}}{(a+i\,t)^x}\,\mathrm{d}t = 0 \tag{2.91b}$$

Proof. Using the Euler relations, see page 15:

$$\cos(b\,t) = \frac{e^{i\,b\,t}+e^{-i\,b\,t}}{2}, \quad \sin(b\,t) = \frac{e^{i\,b\,t}-e^{-i\,b\,t}}{2\,i},$$

in (2.10a) and (2.10b), we obtain the system:

$$\begin{cases} \frac{1}{2}\,(A+B) = \frac{1}{b^x}\,\Gamma(x)\,\cos\frac{\pi\,x}{2}\,, \\ \frac{1}{2\,i}\,(A-B) = \frac{1}{b^x}\,\Gamma(x)\,\sin\frac{\pi\,x}{2}\,, \end{cases}$$

where

$$A = \int_0^\infty t^{x-1} e^{i\,b\,t} \mathrm{d}t, \quad B = \int_0^\infty t^{x-1} e^{-i\,b\,t} \mathrm{d}t\,.$$

We can thus solve obtaining:

$$A = \int_0^\infty t^{x-1} e^{i\,b\,t} \mathrm{d}t = \frac{1}{b^x}\,\Gamma(x)\left(\cos\frac{\pi x}{2} + i\,\sin\frac{\pi x}{2}\right) = \frac{1}{b^x}\,\Gamma(x)\,e^{i\,\frac{\pi\,x}{2}}\,,$$
$$B = \int_0^\infty t^{x-1} e^{-i\,b\,t} \mathrm{d}t = \frac{1}{b^x}\,\Gamma(x)\left(\cos\frac{\pi x}{2} - i\,\sin\frac{\pi x}{2}\right) = \frac{1}{b^x}\,\Gamma(x)\,e^{-i\,\frac{\pi\,x}{2}}\,.$$

Now remember that $i^\alpha = e^{i\,\alpha\,\frac{\pi}{2}}$ and $(-i)^\alpha = e^{-i\,\alpha\,\frac{\pi}{2}}$ we deduce from the two equalities just proved:

$$\int_0^\infty (i\,t)^{x-1} e^{i\,b\,t} \mathrm{d}t = \frac{1}{b^x}\,\Gamma(x)\,e^{i\,\frac{\pi\,x}{2}}\,i^{x-1} = \frac{1}{b^x}\,\Gamma(x)\,e^{i\,\frac{\pi\,x}{2}}\,e^{i\,(x-1)\,\frac{\pi}{2}}$$
$$= -\frac{1}{b^x}\,\Gamma(x)\,i\,e^{i\,\pi\,x} = \frac{1}{i\,b^x}\,\Gamma(x)\,e^{i\,\pi\,x} \tag{1A}$$

$$\int_0^\infty (-i\,t)^{x-1} e^{-i\,b\,t} \mathrm{d}t = \frac{1}{b^x}\,\Gamma(x)\,e^{-i\,\frac{\pi\,x}{2}}\,(-i)^{x-1} = \frac{1}{b^x}\,\Gamma(x)\,e^{-i\,\frac{\pi\,x}{2}}\,e^{-i\,(x-1)\,\frac{\pi}{2}}$$
$$= \frac{1}{b^x}\,\Gamma(x)\,i\,e^{-i\pi\,x} = -\frac{1}{i\,b^x}\,\Gamma(x)\,e^{-i\pi\,x} \tag{1B}$$

$$\int_0^\infty (-i\,t)^{x-1} e^{i\,b\,t} \mathrm{d}t = \frac{1}{b^x}\,\Gamma(x)\,e^{i\,\frac{\pi\,x}{2}}\,(-i)^{x-1}$$
$$= \frac{1}{b^x}\,\Gamma(x)\,e^{i\,\frac{\pi\,x}{2}}\,e^{-i\,(x-1)\,\frac{\pi}{2}} = \frac{i}{b^x}\,\Gamma(x) \tag{2A}$$

$$\int_0^\infty (i\,t)^{x-1} e^{-i\,b\,t} \mathrm{d}t = \frac{1}{b^x}\,\Gamma(x)\,e^{-i\,\frac{\pi\,x}{2}}\,i^{x-1}$$
$$= \frac{1}{b^x}\,\Gamma(x)\,e^{-i\,\frac{\pi\,x}{2}}\,e^{i\,(x-1)\,\frac{\pi}{2}} = -\frac{i}{b^x}\,\Gamma(x) \tag{2B}$$

From the equalities (1A),…, (2B), we deduce two further useful equalities:

$$\int_{-\infty}^\infty (i\,t)^{x-1} e^{i\,b\,t} \mathrm{d}t = \frac{2}{b^x}\,\Gamma(x)\,\sin(\pi\,x) \tag{2.92}$$
$$\int_{-\infty}^\infty (i\,t)^{x-1} e^{-i\,b\,t} \mathrm{d}t = 0 \tag{2.93}$$

which are both correct for real and positive b and for $0 < x < 1$. Splitting the integration domain on the left-hand side of (2.92) and changing the variable as $t = -s$ in the negative branch, obtaining:

$$\begin{aligned}
\int_{-\infty}^{\infty} (i\,t)^{x-1} e^{i\,b\,t} \mathrm{d}t &= \int_{-\infty}^{0} (i\,t)^{x-1} e^{i\,b\,t} \mathrm{d}t + \int_{0}^{\infty} (i\,t)^{x-1} e^{i\,b\,t} \mathrm{d}t \\
&= -\int_{\infty}^{0} (-i\,s)^{x-1} e^{-i\,b\,s} \mathrm{d}s + \int_{0}^{\infty} (i\,t)^{x-1} e^{i\,b\,t} \mathrm{d}t \\
&= \int_{0}^{\infty} (-i\,s)^{x-1} e^{-i\,b\,s} \mathrm{d}s + \int_{0}^{\infty} (i\,t)^{x-1} e^{i\,b\,t} \mathrm{d}t \\
&= -\frac{1}{i\,b^x}\,\Gamma(x)\, e^{-i\pi\,x} + \frac{1}{i\,b^x}\,\Gamma(x)\, e^{i\,\pi\,x} \qquad (*) \\
&= \frac{1}{i\,b^x}\,\Gamma(x)\,\left(e^{i\,\pi\,x} - e^{-i\pi\,x}\right) \\
&= \frac{2}{b^x}\,\Gamma(x)\,\sin(\pi\,x)
\end{aligned}$$

Which shows (2.92) (notice that $(*)$ follows from (1B) and (1A)). Identity (2.93) follows by a similar argument using (2B) and (2A). In order to give the two integrals (2.92) and (2.93) a somewhat more convenient form, one set $t = -a\,i + z$ being a a constant, hence (2.92) and (2.93) are transformed in

$$\int_{a\,i-\infty}^{a\,i+\infty} (a + i\,z)^{x-1} e^{i\,b\,z} \mathrm{d}z = \frac{2}{b^x e^{a\,b}}\,\Gamma(x)\,\sin(\pi\,x) \tag{2.94}$$

$$\int_{a\,i-\infty}^{a\,i+\infty} (a + i\,z)^{x-1} e^{-i\,b\,z} \mathrm{d}z = 0 \tag{2.95}$$

We still have to change the integration path. For this purpose, we use K to denote a large positive number, while we set: $a\,i = \alpha + i\beta$ where α and β are real; further denote $\mathcal{R}$ the perimeter of the rectangle with the vertices $\pm K$ and $\pm K + i\,\beta$; under these conditions the Cauchy's integral theorem yields:

$$\int_{\mathcal{R}} (a + i\,z)^{x-1} e^{\pm i\,b\,z} \mathrm{d}z = 0\,.$$

Whence, running along the four sides of the rectangle, we obtain, if $f(z) = (a + i\,z)^{x-1} e^{\pm i\,b\,z}$:

$$\int_{-K}^{-K+i\,\beta} f(z)\mathrm{d}z + \int_{-K+i\,\beta}^{K+i\,\beta} f(z)\mathrm{d}z + \int_{K+i\,\beta}^{K} f(z)\mathrm{d}z = \int_{-K}^{K} f(z)\mathrm{d}z\,.$$

If the positive number $K \to +\infty$, the first and last integral on the left-hand side vanishes and, from (2.94) and (2.95), we infer

$$\int_{-\infty}^{+\infty} (a + i\,z)^{x-1} e^{i\,b\,z} \mathrm{d}z = \frac{2}{b^x e^{a\,b}}\,\Gamma(x)\,\sin(\pi\,x),$$

$$\int_{-\infty}^{+\infty} (a + i\,z)^{x-1} e^{-i\,b\,z} \mathrm{d}z = 0.$$

Finally replacing x with $1-x$ recalling Euler's reflection formula (2.32) yields the desired integral representations (2.91a) and (2.91b). □

Cauchy calculated the next two integrals in [30], starting from the two integrals (2.91a) and (2.91b) just presented.

Theorem. If $x,\, y > 0$ and $a,\, b > 0$ then

$$\int_{-\infty}^{\infty} \frac{\mathrm{d}u}{(a+i\,u)^x (b+i\,u)^y} = 0\,. \tag{2.96}$$

Moreover if $x+y>1$ then

$$\int_{-\infty}^{\infty} \frac{\mathrm{d}u}{(a+i\,u)^x (b-i\,u)^y} = \frac{2\pi}{(a+b)^{x+y-1}}\,\frac{\Gamma(x+y-1)}{\Gamma(x)\,\Gamma(y)}\,. \tag{2.97}$$

Proof. Recalling proposition 2.2 we have, for $a,\, x>0$

$$\int_0^{\infty} e^{-(a+i\,u)t} t^{x-1}\mathrm{d}t = \frac{1}{(a+i\,u)^x}\,\Gamma(x)\,. \tag{2.98}$$

We use (2.98) in computing both integrals (2.96) and (2.97):

$$\begin{aligned}
\int_{-\infty}^{\infty} \frac{\mathrm{d}u}{(a+i\,u)^x (b\pm i\,u)^y} &= \frac{1}{\Gamma(x)} \int_{-\infty}^{\infty} \left(\int_0^{\infty} e^{-(a+i\,u)t} t^{x-1}\mathrm{d}t \right) \frac{\mathrm{d}u}{(b\pm i\,u)^y} \\
&= \frac{1}{\Gamma(x)} \int_{-\infty}^{\infty} \left(\int_0^{\infty} e^{-at} t^{x-1}\mathrm{d}t \right) \frac{e^{-i\,ut}}{(b\pm i\,u)^y}\mathrm{d}u \\
&= \frac{1}{\Gamma(x)} \int_0^{\infty} e^{-at} t^{x-1}\mathrm{d}t \left(\int_{-\infty}^{\infty} \frac{e^{-i\,ut}}{(b\pm i\,u)^y}\mathrm{d}u \right)
\end{aligned} \tag{2.99}$$

Notice that assumptions about the sign of the parameters allow the order of integration to be exchanged. If in (∗) we choose the + sign then (2.96) follows from (2.91b), while taking the − sign in (2.99) equation (2.96) yields:

$$\begin{aligned}
\int_{-\infty}^{\infty} \frac{\mathrm{d}u}{(a+i\,u)^x (b-i\,u)^y} &= \frac{1}{\Gamma(x)} \int_0^{\infty} e^{-at} t^{x-1}\mathrm{d}t \left(\int_{-\infty}^{\infty} \frac{e^{-i\,ut}}{(b-i\,u)^y}\mathrm{d}u \right) \\
&= \frac{1}{\Gamma(x)} \int_0^{\infty} e^{-at} t^{x-1}\mathrm{d}t \left(\int_{-\infty}^{\infty} \frac{e^{i\,ut}}{(b+i\,u)^y}\mathrm{d}u \right) && \text{(a)} \\
&= \frac{1}{\Gamma(x)} \int_0^{\infty} e^{-at} t^{x-1}\, \frac{2\pi e^{-bt} t^{y-1}}{\Gamma(y)}\mathrm{d}t && \text{(b)} \\
&= \frac{2\pi}{\Gamma(x)\Gamma(y)} \int_0^{\infty} e^{-at} t^{x-1}\, e^{-bt} t^{y-1}\mathrm{d}t \\
&= \frac{2\pi}{\Gamma(x)\Gamma(y)} \int_0^{\infty} t^{x+y-2} e^{-at-bt}\,\mathrm{d}t
\end{aligned}$$

The last integration leads to (2.97). Notice that in (a), we changed the sign at the variable of integration, and in (b), we used the formula (2.91a). □

Corollary 2.21. For x, $y > 0$ and such that $x + y > 1$ we have

$$\int_0^{\frac{\pi}{2}} \cos(y\,s)\,\cos^{x-1} s\,\mathrm{d}s = \frac{\pi\,\Gamma(x)}{2^x\,\Gamma\left(\frac{x+y+1}{2}\right)\Gamma\left(\frac{x-y+1}{2}\right)}\,. \tag{2.100}$$

Formula (2.100) is due to Cauchy see [30, page 302, equation 124]

Proof. We write (2.91a) in the particular case $b = a = 1$:

$$\int_{-\infty}^{\infty} \frac{\mathrm{d}u}{(1+i\,u)^x(1-i\,u)^y} = \frac{\pi}{2^{x+y-2}}\,\frac{\Gamma(x+y-1)}{\Gamma(x)\,\Gamma(y)}\,. \tag{2.101}$$

Using the polar form, we have

$$(1+i\,u)^x(1-i\,u)^y = (1+u^2)^{\frac{x+y}{2}}\,e^{i\,(x-y)\,\arctan u};$$

we can rewrite (2.101) as

$$\int_{-\infty}^{\infty} \frac{e^{-i\,(x-y)\,\arctan u}}{(1+u^2)^{\frac{x+y}{2}}}\mathrm{d}u = \frac{\pi}{2^{x+y-2}}\,\frac{\Gamma(x+y-1)}{\Gamma(x)\,\Gamma(y)}\,. \tag{2.102}$$

Since the right-hand side of (2.102) is real, the imaginary part of the integral vanishes; hence, considering the real part, we infer

$$\int_{-\infty}^{\infty} \frac{\cos\left((x-y)\,\arctan u\right)}{(1+u^2)^{\frac{x+y}{2}}}\mathrm{d}u = \frac{\pi}{2^{x+y-2}}\,\frac{\Gamma(x+y-1)}{\Gamma(x)\,\Gamma(y)}\,. \tag{2.102a}$$

The integrand on the left-hand side of (2.102a) is an even function so that

$$\int_0^{\infty} \frac{\cos\left((x-y)\,\arctan u\right)}{(1+u^2)^{\frac{x+y}{2}}}\mathrm{d}u = \frac{\pi}{2^{x+y-1}}\,\frac{\Gamma(x+y-1)}{\Gamma(x)\,\Gamma(y)}\,. \tag{2.102b}$$

Changing variable as $\arctan u = s$ yields:

$$\int_0^{\frac{\pi}{2}} \cos\left((x-y)\,s\right)\cos^{x+y-2} s\,\mathrm{d}s = \frac{\pi}{2^{x+y-1}}\,\frac{\Gamma(x+y-1)}{\Gamma(x)\,\Gamma(y)}\,. \tag{2.102c}$$

Formula (2.102c) is, indeed, equivalent to our statement (2.100), this is seen solving the system:

$$\begin{cases} x - y = \hat{y}, \\ x + y - 1 = \hat{x}, \end{cases} \implies \begin{cases} x = \dfrac{1}{2}\left(\hat{x}+\hat{y}+1\right), \\ y = \dfrac{1}{2}\left(\hat{x}-\hat{y}+1\right). \end{cases}$$

□

Corollary 2.21 provides some nice particular occurrences; we leave to the Reader the computational details.

Remark. If in (2.100) we take $y = n$ and $x = n+1$ we obtain

$$\int_0^{\frac{\pi}{2}} \cos(n\,x)\,\cos^n x\,\mathrm{d}x = \frac{\pi}{2^{n+1}}. \tag{2.100a}$$

If in (2.100) we take $y = n$ and $x = n-1$ we obtain

$$\int_0^{\frac{\pi}{2}} \cos(n\,x)\,\cos^{n-1} x\,\mathrm{d}x = \frac{2^n}{n\binom{2n}{n}}. \tag{2.100c}$$

If in (2.100) we take $y = n-1$ and $x = n+1$ we obtain

$$\int_0^{\frac{\pi}{2}} \cos((n-1)\,x)\,\cos^n x\,\mathrm{d}x = \frac{2^n}{\binom{2n}{n}}. \tag{2.100b}$$

Integral (2.100) also allows the computation of two interesting integrals.

Corollary. For $x,\, y > 0$ and such that $x + y > 1$ we have

$$\int_0^{\pi} \cos(y\,s)\,\sin^{x-1} s\,\mathrm{d}s = \frac{\pi\,\Gamma(x)\,\cos\frac{\pi\,y}{2}}{2^{x-1}\,\Gamma\left(\frac{x+y+1}{2}\right)\,\Gamma\left(\frac{x-y+1}{2}\right)}, \tag{2.103a}$$

$$\int_0^{\pi} \sin(y\,s)\,\sin^{x-1} s\,\mathrm{d}s = \frac{\pi\,\Gamma(x)\,\sin\frac{\pi\,y}{2}}{2^{x-1}\,\Gamma\left(\frac{x+y+1}{2}\right)\,\Gamma\left(\frac{x-y+1}{2}\right)}. \tag{2.103b}$$

Proof. Due to the symmetries of the cosine and sine functions, we have that

$$\int_{-\frac{\pi}{2}}^{\frac{\pi}{2}} \cos(y\,s)\,\cos^{x-1} s\,\mathrm{d}s = 2\int_0^{\frac{\pi}{2}} \cos(y\,s)\,\cos^{x-1} s\,\mathrm{d}s \tag{a}$$

$$\int_{-\frac{\pi}{2}}^{\frac{\pi}{2}} \sin(y\,s)\,\cos^{x-1} s\,\mathrm{d}s = 0. \tag{b}$$

Multiply the formulas (a) and (b) by $\cos\frac{\pi\,y}{2}$ and $\sin\frac{\pi\,y}{2}$ respectively and then add side by side, obtaining, after using the trigonometric addition cosine formula and recalling Cauchy integral (2.100):

$$\int_{-\frac{\pi}{2}}^{\frac{\pi}{2}} \cos^{x-1} s\,\cos\left[y\left(\tfrac{\pi}{2} - s\right)\right]\,\mathrm{d}s = \frac{\pi\,\Gamma(x)\,\cos\frac{\pi\,y}{2}}{2^{x-1}\,\Gamma\left(\frac{x+y+1}{2}\right)\,\Gamma\left(\frac{x-y+1}{2}\right)}\,.$$

Now we change the variable on the left-hand side as $s + \frac{\pi}{2} = u$ getting:

$$\int_0^{\pi} \cos^{x-1}\left(u - \tfrac{\pi}{2}\right)\,\cos(y\,u)\,\mathrm{d}s = \frac{\pi\,\Gamma(x)\,\cos\frac{\pi\,y}{2}}{2^{x-1}\,\Gamma\left(\frac{x+y+1}{2}\right)\,\Gamma\left(\frac{x-y+1}{2}\right)}\,.$$

Hence (2.103a) follows from the property $\cos\left(u - \frac{\pi}{2}\right) = \sin u$. The formula (2.103b) follows through similar reasoning, but by adding the two equations (a) and (b), previously multiplied by $\sin\frac{\pi\,y}{2}$ and $\cos\frac{\pi\,y}{2}$ respectively. Completing the proof is left to the Reader. □

Chapter 3

Digamma, trigamma, and polygamma

To complete the study of Eulerian functions, in this chapter, we deal with the logarithmic derivative of gamma, known as digamma, which is denoted by the Greek letter ψ. This function allows the representation of harmonic numbers, which are the partial sums of the harmonic series, and has several remarkable integral representations. Theorems 3.7 and 3.9 due to Binet are remarkable since they represent the function $\ln\Gamma(z)$. As we shall see, the derivatives of digamma are, in turn, related to generalised harmonic numbers, e.g. $\psi_1 = \psi'$ is associated with the sum of the reciprocals of the squares and thus to the Basel problem, as stated in formula (3.42).

3.1 Digamma

The function digamma, denoted by ψ, is the logarithmic derivative of gamma hence defined for $x \neq 0, -1, -2, \dots$ as:

$$\psi(x) = \frac{\Gamma'(x)}{\Gamma(x)}.$$

The simplest way to obtain an analytical representation is through the Weierstrass Theorem 2.19, providing the infinite product representation for gamma.

Theorem 3.1. For any $x \neq 0, -1, -2, \dots$ the representation formula holds:

$$\psi(x) = -\gamma + \sum_{n=1}^{\infty}\left(\frac{1}{n} - \frac{1}{n-1+x}\right). \tag{3.1}$$

Proof. Calculating the logarithm of the two sides of (2.64), the infinite product turns into an infinite series, and so we get

$$-\ln\Gamma(x) = \ln x + \gamma\, x + \sum_{n=1}^{\infty}\left(\ln\left(1+\frac{x}{n}\right) - \frac{x}{n}\right). \tag{3.2}$$

DOI: 10.1201/9781003171546-3

Differentiating both sides of (3.2), after changing the sign we find:

$$\psi(x) = \frac{\Gamma'(x)}{\Gamma(x)} = -\gamma - \frac{1}{x} + \sum_{n=1}^{\infty}\left(\frac{1}{n} - \frac{1}{n+x}\right).$$

Bringing x into the summation yields the thesis (3.1). □

From (3.1), we immediately deduce the recursive relation that characterises $\psi(x)$.

Corollary. For any $x \neq 0, -1, -2, \ldots$ we have

$$\psi(x+1) = \psi(x) + \frac{1}{x}. \tag{3.3}$$

Proof. Since the following relation holds

$$\psi(x+1) - \psi(x) = \sum_{n=1}^{\infty}\left(\frac{1}{n-1+x} - \frac{1}{n+x}\right)$$

observed that the left-hand side series telescopes:

$$\sum_{n=1}^{\infty}\left(\frac{1}{n-1+x} - \frac{1}{n+x}\right) = \left(\frac{1}{x} - \frac{1}{x+1}\right) + \left(\frac{1}{x+1} - \frac{1}{x+2}\right)\cdots$$

Hence (3.3) follows. □

Remark. Formula (3.3) is generalised by inductive reasoning: if $n \in \mathbb{N}$ then

$$\psi(x+n) = \psi(x) + \frac{1}{x} + \frac{1}{x+1} + \cdots + \frac{1}{x+n-1} \tag{3.4}$$

Remark. Definition (3.1) implies that $\psi(1) = \Gamma'(1) = -\gamma$ thus (2.5a) yields:

$$-\gamma = \psi(1) = \Gamma'(1) = \int_0^{+\infty} e^{-t}\ln t\,\mathrm{d}t. \tag{3.5}$$

Remark. The values of $\psi(x)$ at positive integers are related to the harmonic numbers since (3.4) for $x = n \in \mathbb{N}$ provides

$$\psi(n+1) = -\gamma + 1 + \frac{1}{2} + \frac{1}{3} + \cdots + \frac{1}{n} = -\gamma + H_n. \tag{3.6}$$

Reflection formula for ψ

As a consequence of the reflection formula (2.32), we infer

Theorem. For any $x \in]0,1[$ we have

$$\psi(x) - \psi(1-x) = -\frac{\pi}{\tan \pi x} \tag{3.7}$$

Proof. The statement is obtained by direct, straightforward computation:

$$\begin{aligned}\psi(x)-\psi(1-x) &= \frac{\mathrm{d}}{\mathrm{d}x}\ln\Gamma(x)+\frac{\mathrm{d}}{\mathrm{d}x}\ln\Gamma(1-x)\\ &= \frac{\mathrm{d}}{\mathrm{d}x}\ln\left[\Gamma(x)\Gamma(1-x)\right]\\ &= \frac{\mathrm{d}}{\mathrm{d}x}\ln\frac{\pi}{\sin\pi x}=\frac{\sin\pi x}{\pi}\,\frac{-\pi^2\cos\pi x}{(\sin\pi x)^2}\\ &= -\frac{\pi}{\tan\pi x}.\end{aligned}$$

□

Multiplication formula for ψ

As a consequence of the duplication formula, (2.50), we can deduce the duplication formula for ψ.

Theorem.

$$\psi(2x)=\ln 2+\frac{1}{2}\left(\psi(x)+\psi\left(x+\frac{1}{2}\right)\right). \tag{3.8}$$

Proof. We calculate the logarithm of both sides of (2.50):

$$\begin{aligned}\ln\Gamma(2x) &= \ln\left(\frac{2^{2x-1}}{\sqrt{\pi}}\Gamma(x)\Gamma\left(x+\frac{1}{2}\right)\right)\\ &= \ln 2^{2x-1}-\ln\sqrt{\pi}+\ln\Gamma(x)+\ln\Gamma\left(x+\frac{1}{2}\right).\end{aligned}$$

Then, differentiating, we obtain

$$2\psi(2x)=2\ln 2+\psi(x)+\psi\left(x+\frac{1}{2}\right),$$

which concludes the demonstration of (3.8). □

The Digamma multiplication formula follows from the Gamma multiplication formula (2.49). We leave the proof to the reader. Here, we limit ourselves to its statement:

$$\psi(n\,x)=\ln n+\frac{1}{n}\sum_{k=0}^{n-1}\psi\left(x+\frac{k}{n}\right).$$

Integral representations

There are several integral representation formulas. We begin by presenting two of the simplest:

Proposition 3.2. For $x > 0$, the following formulas hold:

$$\psi(x) = -\gamma + \int_0^1 \frac{1-u^{x-1}}{1-u} \mathrm{d}u, \tag{3.9}$$

$$\psi(x) = -\gamma + \int_0^\infty \frac{e^{-t}-e^{-xt}}{1-e^{-t}} \mathrm{d}t. \tag{3.10}$$

Proof. We observe that the integrals in (3.9) and (3.10) are closely related: it is enough to start from (3.10) and change the variable by placing $e^{-t} = u$. It will then be sufficient to show (3.9) to prove our assertions; at this aim we have to show:

$$\sum_{n=1}^\infty \left(\frac{1}{n} - \frac{1}{n-1+x}\right) = \int_0^1 \frac{1-u^{x-1}}{1-u} \mathrm{d}u. \tag{3.11}$$

In fact, on the left-hand side integral in (3.11), we expand $(1-u)^{-1}$ in a geometric series and exchange the series with the integral, a permissible step thanks to the monotone convergence theorem, obtaining (3.11), indeed:

$$\begin{aligned}\int_0^1 \frac{1-u^{x-1}}{1-u} \mathrm{d}u &= \sum_{n=1}^\infty \int_0^1 u^{n-1}(1-u^{x-1})\mathrm{d}u = \sum_{n=1}^\infty \int_0^1 (u^{n-1} - u^{n+x-2})\mathrm{d}u \\ &= \sum_{n=1}^\infty \left(\frac{1}{n} - \frac{1}{n+x-1}\right).\end{aligned}$$

□

Remark. From (3.9) we deduce the integration formula, where a, $b > 0$

$$\int_0^1 \frac{u^a - u^b}{1-u} \, \mathrm{d}u = \psi(b+1) - \psi(a+1). \tag{3.12}$$

When, in particular, the parameters a and b assume rational values, the integral (3.12) is evaluated in terms of elementary functions, and this fact is related to a theorem due to Gauss, see equation (3.39).

Remark. (3.6) and (3.9) when $x = n+1 \in \mathbb{N}$ provide the integral representation of the harmonic numbers due to Euler:

$$H_n = \int_0^1 \frac{1-u^n}{1-u} \mathrm{d}u\,. \tag{3.13}$$

From the formula for the representation of the harmonic numbers (3.13), we deduce a further relation of integral representation of digamma, in which the Euler-Mascheroni constant does not appear. We derive an integral representation of γ to obtain this result, which is of great interest.

Proposition 3.3. If γ is defined by (2.61), then

$$\gamma = \int_0^1 \left(\frac{1}{1-x} + \frac{1}{\ln x} \right) \mathrm{d}x \,. \tag{3.14}$$

Proof. We begin with the identity:

$$\ln n = \int_0^1 \frac{x^{n-1} - 1}{\ln x} \, \mathrm{d}x \,, \tag{3.15}$$

which is obtained by differentiating with respect to n the parametric integral on the right-hand side in (3.15), so recalling equation (2.61), we see that

$$\begin{aligned}
\gamma &= \lim_{n\to+\infty} \left(\sum_{k=1}^{n} \frac{1}{k} - \ln n \right) \\
&= \lim_{n\to+\infty} \left(\int_0^1 \frac{1-x^n}{1-x} \mathrm{d}x - \int_0^1 \frac{x^{n-1}-1}{\ln x} \, \mathrm{d}x \right) \\
&= \lim_{n\to+\infty} \int_0^1 \left(\frac{1-x^n}{1-x} - \frac{x^{n-1}-1}{\ln x} \right) \mathrm{d}x.
\end{aligned}$$

The formula (3.14) now follows from the evaluation of the limit. □

We are now assembling equation (3.14) with the basic integral representation (3.9). We arrive at further representations of ψ, due to Gauss, which will be fundamental in calculating an important integral studied by Binet, which we will cover later on page 116.

Theorem. For $x > 0$, the integral representations hold:

$$\psi(x) = \int_0^\infty \left(\frac{e^{-t}}{t} - \frac{e^{-xt}}{1-e^{-t}} \right) \mathrm{d}t \tag{3.16}$$

$$\psi(x) = -\int_0^1 \left(\frac{1}{\ln u} + \frac{u^{x-1}}{1-u} \right) \mathrm{d}u \tag{3.17}$$

Proof. From (3.10) replacing γ with (3.14) we obtain

$$\psi(x) = -\int_0^1 \left(\frac{1}{1-u} + \frac{1}{\ln u} \right) \mathrm{d}u + \int_0^\infty \frac{e^{-t} - e^{-xt}}{1-e^{-t}} \mathrm{d}t \,.$$

Changing the variable on the first integral on the right-hand side, $-\ln u = t$ yields:

$$\begin{aligned}
\psi(x) &= -\int_0^\infty \left(\frac{e^{-t}}{1-e^{-t}} - \frac{e^{-t}}{t} \right) \mathrm{d}t + \int_0^\infty \frac{e^{-t} - e^{-xt}}{1-e^{-t}} \mathrm{d}t \\
&= \int_0^\infty \left(\frac{e^{-t}}{t} - \frac{e^{-xt}}{1-e^{-t}} \right) \mathrm{d}t.
\end{aligned}$$

Showing (3.16), which is due to Gauss. Similarly, recalling, again, (3.9) and, replacing γ with (3.14) we obtain

$$\psi(x) = -\int_0^1 \left(\frac{1}{1-u} + \frac{1}{\ln u} \right) du + \int_0^1 \frac{1-u^{x-1}}{1-u} du \,.$$

Hence (3.17) yields adding the integrals on the right-hand side. □

From (3.16), a further representation formula is deduced, which is useful in demonstrating an important estimate for the digamma Theorem 3.6 below.

Proposition 3.4. If $x > 0$ then

$$\psi(x) = \ln x + \int_0^\infty \left(\frac{1}{t} - \frac{1}{1-e^{-t}} \right) e^{-t\,x} \, dt \,. \tag{3.18}$$

Proof. We proceed in a manner very similar to that followed in the demonstration of proposition 3.3 using a specific (Frullani type) representation of the logarithmic function; in this case, instead of the formula (3.15), we use

$$\ln x = \int_0^\infty \frac{e^{-t} - e^{-x\,t}}{t} \, dt \,. \tag{3.19}$$

Formula (3.19) follows from

$$\frac{1}{u} = \int_0^\infty e^{-u\,t} \, dt \,. \tag{$\oplus$}$$

Then, integrating ($\oplus$) in $[1, x]$ and switching the order of integration, we obtain

$$\ln x = \int_1^x \frac{du}{u} = \int_1^x \left(\int_0^\infty e^{-u\,t} \, dt \right) du = \int_0^\infty \left(\int_1^x e^{-u\,t} \, du \right) dt \,.$$

Thus, (3.19) follows from a straightforward computation. Now, in the integrand of (3.16) we add and remove $t^{-1} e^{-x\,t}$ so that, recalling (3.19), we arrive at (3.18)

$$\begin{aligned}
\psi(x) &= \int_0^\infty \left(\frac{e^{-t}}{t} - \frac{e^{-xt}}{1-e^{-t}} \right) dt = \int_0^\infty \left(\frac{e^{-t} - e^{-x\,t}}{t} + \frac{e^{-x\,t}}{t} - \frac{e^{-xt}}{1-e^{-t}} \right) dt \\
&= \int_0^\infty \left(\frac{e^{-t} - e^{-x\,t}}{t} \right) dt + \int_0^\infty \left(\frac{1}{t} - \frac{1}{1-e^{-t}} \right) e^{-x\,t} \, dt \\
&= \ln x + \int_0^\infty \left(\frac{1}{t} - \frac{1}{1-e^{-t}} \right) e^{-x\,t} \, dt \,.
\end{aligned}$$

□

We conclude the review of ψ integral representations by illustrating the one provided by Dirichlet, which is interesting because it uses a technique that is very common when trying to establish identities, that of calculating the same quantity, in this case, a double integral, using two different methods. We shall again use the representation of the logarithm introduced in (3.19), the derivative of gamma computed in (2.5a), and formula (2.7).

Proposition 3.5. If $x > 0$ then

$$\psi(x) = \int_0^\infty \frac{1}{z}\left(e^{-z} - \frac{1}{(1+z)^x}\right) \mathrm{d}z. \tag{3.20}$$

Proof. Consider the double integral:

$$D(x) = \int_0^\infty \int_0^\infty s^{x-1} \frac{e^{-s-z} - e^{-s\,(z+1)}}{z} \,\mathrm{d}s\,\mathrm{d}z. \tag{3.21}$$

Integrating (3.21) starting from z we can invoke (3.19) and (2.5a) as follows:

$$\begin{aligned} D(x) &= \int_0^\infty s^{x-1} e^{-s} \left(\int_0^\infty \frac{e^{-z} - e^{-s\,z}}{z} \,\mathrm{d}z \right) \mathrm{d}s \\ &= \int_0^\infty s^{x-1} e^{-s} \ln s \,\mathrm{d}s = \Gamma'(x)\,. \end{aligned}$$

On the other side, reverting the order of integration in (3.21), we find out:

$$\begin{aligned} D(x) &= \int_0^\infty \frac{1}{z} \left(\int_0^\infty s^{x-1} \left(e^{-s-z} - e^{-s(z+1)} \right) \mathrm{d}s \right) \mathrm{d}z \\ &= \int_0^\infty \frac{1}{z} \left(e^{-z} \int_0^\infty s^{x-1} e^{-s} \mathrm{d}s - \int_0^\infty s^{x-1} e^{-(z+1)s} \mathrm{d}s \right) \mathrm{d}z \\ &= \int_0^\infty \frac{1}{z} \left(e^{-z}\Gamma(x) - \frac{1}{(1+z)^x}\Gamma(x) \right) \mathrm{d}z. \end{aligned}$$

Notice that in the last step, we used the formula (2.7). In conclusion, we have shown that

$$D(x) = \Gamma(x) \int_0^\infty \frac{1}{z}\left(e^{-z} - \frac{1}{(1+z)^x}\right) \mathrm{d}z.$$

The Dirichlet formula (3.20) follows from comparing the two expressions obtained for the double integral $D(x)$. □

Asymptotic behaviour

As mentioned, the formula (3.18) allows the determination of a double estimate from above and below for digamma due to Minc and Sathre [81].

Theorem 3.6. If $x > 1$ then

$$\ln x - \frac{1}{x} < \psi(x) < \ln x - \frac{1}{2x}. \tag{3.22}$$

Proof. The proof presented here refers to the cited article [81]. In view of (3.18) to establish inequality (3.22) we simply have to show that, for $t > 0$:

$$-1 < \frac{1}{t} - \frac{1}{1-e^{-t}} < -\frac{1}{2}. \tag{3.23}$$

Changing the sign, we rewrite (3.23) more conveniently as

$$\frac{1}{2} < \frac{1}{1-e^{-t}} - \frac{1}{t} < 1. \tag{3.24}$$

We focus on the right-hand side of (3.24). Our starting point is the inequality $(1+t)\,e^{-t} < 1$, which immediately follows from the fact, elementarily verifiable with basic advanced calculus methods, that

$$\sup_{t\in\mathbb{R}} (t+1)e^{-t} = 1\,.$$

Observing that $(t+1)e^{-t} < 1 \iff e^{-t} < 1 - te^{-t}$ we have

$$t - 1 + e^{-t} < t - 1 + 1 - te^{-t} = t(1-e^{-t})\,.$$

Now, summing the middle term in (3.24) and using the previous inequality, we see that

$$\frac{1}{1-e^{-t}} - \frac{1}{t} = \frac{t-1+e^{-t}}{t(1-e^{-t})} < \frac{t(1-e^{-t})}{t(1-e^{-t})} = 1\,,$$

which indeed shows the right-hand side inequality in (3.24). To prove the left-hand side inequality in (3.24), we start by observing that for $t > 0$ inequality

$$t - 2 + (t+2)e^{-t} > 0, \tag{$\ominus$}$$

holds. First observed that ($\ominus$) is equivalent to $(t+2)e^{-t} > 2 - t$, and by differentiating both sides of the last inequality, we find

$$-e^{-t}(t+1) > -1 \iff (1+t)e^{-t} < 1,$$

which is, indeed, true, as we already noticed. At this point, the inequality ($\ominus$) follows from the fact, well known from advanced calculus courses, that from the conditions $f(0) \geq (0)$ and $f'(t) > g'(t)$ for each $t > 0$ follows $f(t) > g(t)$ for $t > 0$. Now, from ($\ominus$) it follows that

$$t - 2 + te^{-t} + 2e^{-t} > 0 \iff 2t - 2 + 2e^{-t} > t - te^{-t} = t(1-e^{-t}).$$

Now, switching to reciprocal:

$$\frac{1}{2t-2+2e^{-t}} < \frac{1}{t(1-e^{-t})}. \tag{$\otimes$}$$

Summing again the middle term in (3.24) and using inequality ($\otimes$) we see that

$$\frac{1}{1-e^{-t}} - \frac{1}{t} = \frac{t-1+e^{-t}}{t(1-e^{-t})} > \frac{t-1+e^{-t}}{2t-2+2e^{-t}} = \frac{1}{2}.$$

With this, the proof of (3.23) is complete and, as already noted, the whole theorem is therefore proved. □

Remark. Theorem 3.6 formalises the asymptotic behaviour of digamma, which, on closer inspection, can be understood heuristically by comparing the link between ψ and harmonic numbers, see equations (3.4) and (3.1) and the link between the latter and the Euler–Mascheroni constant equation introduced in (2.61). We also point out that H. Alzer in [2, 3] proved the validity of (3.23) to $x > 0$ and in [57] a refinement of (3.23) is provided.

Binet formulas, for logarithm of Gamma

We present here two formulas due to Binet (1786–1856) concerning determining the logarithm of gamma, using, among others, properties of digamma and the Stirling formula.

Theorem 3.7. For $z > 0$ we have

$$\ln \Gamma(z) = \left(z - \frac{1}{2}\right) \ln z - z + \ln \sqrt{2\pi} + \int_0^\infty \left(\frac{1}{2} - \frac{1}{u} + \frac{1}{e^u - 1}\right) \frac{e^{-zu}}{u} du \quad (3.25)$$

Remark. As a matter of fact, (3.25), as well (3.31) below, could be proved for complex arguments z with a positive real part by replacing the real logarithm, which for this reason and to avoid ambiguity has been indicated with "ln", with the principal value of the complex logarithm "Log".

Proof. Let us start by studying the parametric integral that appears on the right-hand side of (3.25):

$$I(z) = \int_0^\infty \left(\frac{1}{2} - \frac{1}{u} + \frac{1}{e^u - 1}\right) \frac{e^{-zu}}{u} du.$$

Despite its appearance, the integrand is bounded as $u \to 0^+$, whereas the convergence is exponential at infinity. Differentiation with respect to z is then permissible, leading to

$$\begin{aligned} I'(z) &= -\int_0^\infty \left(\frac{1}{2} - \frac{1}{u} + \frac{1}{e^u - 1}\right) e^{-zu} \, du \\ &= -\int_0^\infty \frac{e^{-zu}}{2} du - \int_0^\infty \left(\frac{1}{e^u - 1} - \frac{1}{u}\right) e^{-zu} \, du \\ &= -\frac{1}{2z} - \int_0^\infty \left(\frac{1}{e^u - 1} - \frac{1}{u}\right) e^{-zu} \, du \\ &= -\frac{1}{2z} - J(z), \end{aligned} \quad (3.26)$$

where

$$J(z) = \int_0^\infty \left(\frac{1}{e^u - 1} - \frac{1}{u}\right) e^{-zu} \, du \, .$$

Now, after changing the variable in $J(z)$ by setting $\bar{u} = e^{-u}$, we arrive at, after some computations, and use again u as the integration variable

$$J(z) = \int_0^1 \left(\frac{u^z}{1-u} + \frac{u^{z-1}}{\ln u} \right) du \,.$$

Next, we rewrite the integrand by adding and subtracting 1 to the numerator of the second addend:

$$\begin{aligned} J(z) &= \int_0^1 \left(\frac{u^z}{1-u} + \frac{u^{z-1} - 1 + 1}{\ln u} \right) du \\ &= \int_0^1 \left(\frac{u^z}{1-u} + \frac{1}{\ln u} \right) du + \int_0^1 \frac{u^{z-1} - 1}{\ln u} du. \end{aligned} \tag{3.27}$$

Looking on the right-hand side of (3.27), we see that the first addend is expressible using (3.17), while for the second, we invoke (3.15), so that

$$J(z) = \ln z - \psi(1+z). \tag{3.28}$$

Thus, inserting (3.28) into (3.26), we arrive at

$$I'(z) = -\frac{1}{2z} - \ln z + \psi(1+z).$$

Hence, integrating:

$$\begin{aligned} I(z) &= -\frac{1}{2} \ln z - (z \ln z - z) + \ln \Gamma(1+z) + c \\ &= -\left(z + \frac{1}{2} \right) \ln z + z + \ln \Gamma(1+z) + c. \end{aligned} \tag{$\diamond$}$$

Observe that c is an integration constant that is determined by observing that the boundary condition holds:

$$\lim_{z \to +\infty} I(z) = 0.$$

Then, rewrite ($\diamond$) using well-known properties of the logarithm and then the gamma functional property (2.4), as

$$I(z) = \ln \left(\frac{e^z e^c \, \Gamma(z+1)}{z^{z+\frac{1}{2}}} \right) = \ln \left(\frac{e^z z^{\frac{1}{2}} e^c \, \Gamma(z)}{z^z} \right).$$

Now, using the Stirling formula, equation (2.48), we arrive at

$$\lim_{z \to +\infty} I(z) = \lim_{z \to +\infty} \ln \left(\frac{e^z \, z^{\frac{1}{2}} e^c \, \Gamma(z)}{z^z} \right) = \ln(\sqrt{2\pi} \, e^c) = 0,$$

so that $\sqrt{2\pi}\, e^c = 1$ and finally $c = -\ln\sqrt{2\pi}$. Plugging in ($\diamond$), using again the functional property of gamma

$$\begin{aligned} I(z) &= -\left(z+\frac{1}{2}\right)\ln z + z + \ln z + \ln\Gamma(z) - \ln\sqrt{2\pi} \\ &= \left(\frac{1}{2}-z\right)\ln z + z + \ln\Gamma(z) - \ln\sqrt{2\pi}. \end{aligned}$$

Finally, retrieving $\ln\Gamma(z)$ from the last expression, recalling the definition of $I(z)$ we infer

$$\ln\Gamma(z) = \left(z-\frac{1}{2}\right)\ln z - z + \ln\sqrt{2\pi} + \int_0^\infty \left(\frac{1}{2} - \frac{1}{u} + \frac{1}{e^u - 1}\right)\frac{e^{-zu}}{u}\, du\,.$$

This ends the proof of the first Binet formula (3.25). □

With a similar technique, which, because of its importance, we report in full detail, we provide the proof of Binet's second formula for $\ln\Gamma(z)$. First of all, we use a preliminary result, which is indispensable for obtaining Binet's second formula, which essentially allows the calculation of a definite integral using, among other things, the formula for the series development of the hyperbolic cotangent (1.35).

Lemma 3.8. If $a > 0$ is a positive real number, then

$$\int_0^\infty \frac{\sin(a\,x)}{1-e^x}\, dx = \frac{1}{2}\left(\pi\coth(\pi a) - \frac{1}{a}\right). \tag{3.29}$$

Proof. The integral on the left-hand side of (3.29) is related to the McLaurin series of the hyperbolic cotangent; see equation (1.35). Starting from the partial decomposition, we see that

$$\begin{aligned} \frac{a}{n^2+a^2} &= \frac{i}{2}\left(\frac{1}{n+i\,a} - \frac{1}{n-i\,a}\right) \\ &= \frac{i}{2}\left(\int_0^\infty \left(e^{-(n+i\,a)x} - e^{-(n-i\,a)x}\right) dx\right) \\ &= \frac{i}{2}\left(\int_0^\infty \left(-2\,i\sin(a\,x)\,e^{-nx}\right) dx\right) \\ &= \int_0^\infty \sin(a\,x)\,e^{-nx}\, dx. \end{aligned}$$

Summing for n from 1 to ∞, observing the correctness of the exchange between series and integral, and finally using the sum of the geometric series, we obtain

$$\sum_{n=1}^{\infty} \frac{a}{a^2+n^2} = \int_0^\infty \frac{\sin a\,x}{e^x - 1}\, dx\,. \tag{3.30}$$

On the other side, the sum of the series that appears on the left-hand side of (3.30) is evaluated in (1.35), and this shows (3.29). □

Theorem 3.9. For $z > 0$, we have that

$$\ln\Gamma(z) = \left(z - \frac{1}{2}\right)\ln z - z + \frac{1}{2}\ln(2\pi) + 2\int_0^{+\infty} \frac{\arctan\left(\frac{t}{z}\right)}{e^{2\pi t} - 1}\,dt \tag{3.31}$$

Proof. We will employ the integral representation for the inverse tangent

$$\arctan\left(\tfrac{x}{z}\right) = \int_0^{\infty} \frac{\sin(x\,t)\,e^{-z\,t}}{t}\,dt\,. \tag{3.32}$$

To verify (3.32), it is sufficient to differentiate both sides, note that they are equal and then observe that for $x = 0$, the (3.32) is satisfied.

Now consider the integral, looking on the right-hand side of (3.31):

$$I(z) = 2\int_0^{\infty} \frac{\arctan\left(\frac{x}{z}\right)}{e^{2\pi x} - 1}\,dt\,. \tag{3.33}$$

Thus, by inserting (3.32) into (3.33), we can use Fubini's theorem to obtain

$$\begin{aligned} I(z) &= 2\int_0^{\infty} \frac{1}{e^{2\pi x} - 1}\left(\int_0^{\infty} \frac{\sin(x\,t)\,e^{-z\,t}}{t}\,dt\right)dx \\ &= 2\int_0^{\infty} \frac{e^{-z\,t}}{t}\left(\int_0^{\infty} \frac{\sin(x\,t)}{e^{2\pi x} - 1}\,dx\right)dt. \end{aligned} \tag{3.34}$$

To evaluate the inner integral on the right-hand side of (3.34) we use the equation (3.29) showed in Lemma 3.8, immediately after changing variable $2\pi\,x = u$ obtaining:

$$\int_0^{\infty} \frac{\sin(x\,t)}{e^{2\pi x} - 1}\,dx = \frac{1}{2\pi}\int_0^{\infty} \frac{\sin\left(\frac{tu}{2\pi}\right)}{e^u - 1}\,du = \frac{1}{2\,\pi}\left(\frac{1}{2}\pi\coth\left(\frac{t}{2}\right) - \frac{\pi}{t}\right).$$

Consequently, returning to (3.34) we see that

$$\begin{aligned} I(z) &= 2\int_0^{\infty} \frac{e^{-z\,t}}{t}\left(\frac{1}{4}\coth\left(\frac{t}{2}\right) - \frac{1}{2t}\right)dt \\ &= \frac{1}{2}\int_0^{\infty} \frac{e^{-2z\,t}}{t}\left(\coth t - \frac{1}{t}\right)dt. \end{aligned} \tag{3.35}$$

In the last step, we changed the variable by putting $2t = t'$, and for simplicity, we used the variable t without accent again. We now differentiate with respect to z the parametric integral (3.35), obtaining:

$$I'(z) = \int_0^{\infty} e^{-2z\,t}\left(\frac{1}{t} - \coth t\right)dt$$

$$= \int_0^{\infty} e^{-2z\,t}\left(\frac{1}{t} - \frac{1 + e^{-2\,t}}{1 - e^{-2\,t}}\right)dt \tag{3.36a}$$

$$= \frac{1}{2}\int_0^{\infty} e^{-z\,t}\left(\frac{2}{t} - \frac{1 + e^{-t}}{1 - e^{-t}}\right)dt \tag{3.36b}$$

$$= \int_0^{\infty} e^{-z\,t}\left(\frac{1}{t} - \frac{1}{2}\frac{1 + e^{-t}}{1 - e^{-t}}\right)dt\,. \tag{3.36c}$$

Notice that in (3.36a) we express $\coth t$ in exponential terms, in (3.36b) we change variable $2t = t'$ (again omitting the apostrophe) arriving at (3.36c) with a straightforward simplification. The next step at this point is to exploit representation formula (3.18) for ψ proved in proposition 3.4; thus, we can rewrite (3.36c) as follows:

$$\begin{aligned}
I'(z) &= \int_0^\infty e^{-z\,t}\left(\frac{1}{t} - \frac{1}{1-e^{-t}} + \frac{1}{1-e^{-t}} - \frac{1}{2}\frac{1+e^{-t}}{1-e^{-t}}\right) \mathrm{d}t \\
&= \int_0^\infty e^{-z\,t}\left(\frac{1}{t} - \frac{1}{1-e^{-t}}\right) \mathrm{d}t + \int_0^\infty e^{-z\,t}\left(\frac{1}{1-e^{-t}} - \frac{1}{2}\frac{1+e^{-t}}{1-e^{-t}}\right) \mathrm{d}t \\
&= \int_0^\infty e^{-z\,t}\left(\frac{1}{t} - \frac{1}{1-e^{-t}}\right) \mathrm{d}t + \frac{1}{2}\int_0^\infty e^{-z\,t}\,\mathrm{d}t \\
&= \psi(z) - \ln z + \frac{1}{2z}\,. \qquad (3.37)
\end{aligned}$$

Notice that (3.37) is a consequence of (3.18). From this point on, the demonstration is identical to that set out in Theorem 3.7, so we leave it to the reader. □

As pointed out in [19, equation (27) p. 18], we can obtain a further integral representation of ψ, which will find an interesting use in chapter 4 on zeta functions, Theorem 4.6.

Corollary 3.10. If $z > 0$, then

$$\psi(z) = \ln z - \frac{1}{2z} - 2\int_0^\infty \frac{\mathrm{d}t}{(t^2+z^2)(e^{2\pi t}-1)}. \qquad (3.38)$$

Proof. It is enough to differentiate with respect to z on the right-hand side of the formula (3.31):

$$\frac{\mathrm{d}}{\mathrm{d}z}\left(\left(z-\frac{1}{2}\right)\ln z - z - \frac{1}{2}\ln(2\pi) + 2\int_0^{+\infty}\frac{\arctan\left(\frac{t}{z}\right)}{e^{2\pi\,t}-1}\,\mathrm{d}t\right) =$$

$$1 - \frac{1}{2z} + \ln z - \int_0^\infty \frac{2t}{(t^2+z^2)\,(e^{2\pi t}-1)}\,\mathrm{d}t$$

proving (3.38). □

Digamma at rational argument

A celebrated result due to Gauss [52, pp. 155–156], evaluates ψ with rational argument.

Theorem 3.11. For $p,\, q \in \mathbb{N},\, p < q$

$$\psi\left(\tfrac{p}{q}\right) = -\gamma - \ln(2q) - \tfrac{\pi}{2}\cot\left(\tfrac{p}{q}\,\pi\right) + \sum_{k=1}^{\lceil q/2\rceil - 1} \cos\left(\tfrac{2kp\pi}{q}\right)\ln\left(2\sin\left(\tfrac{k\pi}{q}\right)\right) \qquad (3.39)$$

where $\lceil x \rceil$ is the so-called ceiling function, which maps x to the least integer greater than or equal to x:

$$\lceil x \rceil = \max\{y \in \mathbb{Z} : y \geq x\}.$$

A proof, certainly less demanding than the original, is provided in [68], where we refer interested readers. Here, we limit ourselves to calculating $\psi(\frac{1}{2})$ based on (3.39), observe that for this value the sum in (3.39) vanishes, since it is indeed empty hence $\psi(\frac{1}{2}) = -\gamma - \ln 4$.

3.2 Trigamma and polygammas

Definition. The trigamma function, which we will denote by ψ_1, is the first derivative of ψ, so it is defined by the relation

$$\psi_1(x) = \frac{\mathrm{d}}{\mathrm{d}x}\psi(x) = \frac{\Gamma(x)\Gamma''(x) - (\Gamma'(x))^2}{\Gamma^2(x)}.$$

Theorem. For trigamma, we have the series representation:

$$\psi_1(x) = \sum_{n=1}^{\infty} \frac{1}{(n-1+x)^2}, \tag{3.40}$$

and the recursive relation:

$$\psi_1(x+1) = \psi_1(x) - \frac{1}{x^2}. \tag{3.41}$$

Proof. From Theorem 3.1, equation (3.1), we obtain (3.40) differentiating term by term and (3.40) in turn, implies (3.41). □

Remark. The value of ψ_1 in $x = 1$ is that of the harmonic series of order 2 (Basel problem), a value we calculated in (1.30)

$$\psi_1(1) = \sum_{n=1}^{\infty} \frac{1}{n^2} = \frac{\pi^2}{6}. \tag{3.42}$$

The value of ψ_1 in $x = 1/2$ is also computed using (1.30).

Theorem.

$$\psi_1\left(\frac{1}{2}\right) = \frac{\pi^2}{2}.$$

Proof. From (3.40) follows that

$$\psi_1\left(\tfrac{1}{2}\right) = \sum_{n=1}^{\infty} \frac{1}{\left(n - \frac{1}{2}\right)^2} = 4\sum_{n=1}^{\infty} \frac{1}{(2n-1)^2}.$$

The sum of the reciprocals of the squares of the odd numbers is related to the Basel sum (1.30), indeed:

$$\begin{aligned}\frac{\pi^2}{6} &= \sum_{n=1}^{\infty} \frac{1}{n^2} = \sum_{n=1}^{\infty} \frac{1}{(2n)^2} + \sum_{n=1}^{\infty} \frac{1}{(2n-1)^2} \\ &= \frac{1}{4}\sum_{n=1}^{\infty} \frac{1}{n^2} + \sum_{n=1}^{\infty} \frac{1}{(2n-1)^2} = \frac{\pi^2}{24} + \sum_{n=1}^{\infty} \frac{1}{(2n-1)^2}\end{aligned}$$

hence

$$\sum_{n=1}^{\infty} \frac{1}{(2n-1)^2} = \frac{\pi^2}{6} - \frac{\pi^2}{24} = \frac{\pi^2}{8}.$$

□

Remark. From the integral representation (3.9) of Theorem (3.2), deriving term by term, we obtain the integral representation formula for ψ_1:

$$\psi_1(x) = -\int_0^1 \frac{t^{x-1}\ln t}{1-t}\,\mathrm{d}t.$$

Also from reflexion formula (2.32) we infer that for $0 < x < 1$

$$\psi_1(x) + \psi_1(1-x) = \frac{\pi^2}{\sin^2(\pi x)}.$$

Definition. The polygamma function of order $n \in \mathbb{N}$ is the n-th derivative of the digamma function:

$$\psi_n(x) = \frac{\mathrm{d}^n}{\mathrm{d}x^n}\,\psi(x).$$

The domain of definition of $\psi_n(x)$ is the same of ψ i.e. $x \neq 0, -1, -2, \ldots$

Remark. Recalling (3.1) we see that

$$\begin{aligned}\psi_n(x) &= \frac{\mathrm{d}^n}{\mathrm{d}x^n}\left(\sum_{k=1}^{\infty}\left(\frac{1}{k} - \frac{1}{k-1+x}\right)\right) \\ &= -\frac{\mathrm{d}^n}{\mathrm{d}x^n}\left(\sum_{k=0}^{\infty}\frac{1}{k+x}\right) \qquad \text{(a)} \\ &= \sum_{k=0}^{\infty}(-1)^{n+1}\frac{n!}{(k+x)^{n+1}}. \qquad (3.43)\end{aligned}$$

In (a), first, we did not consider the constant term, which the differentiation operation causes to cancel, and then we rescaled the summation index. Finally, (3.43) follows from the derivative of the reciprocal.

The reflection formula can also be formulated in this circumstance as follows.

Theorem. For any $n \in \mathbb{N}$ and for any $0 < x < 1$

$$\psi_n(x) + (-1)^n \psi_n(1-x) = -\pi \frac{\mathrm{d}^n}{\mathrm{d}x^n} \cot(\pi x).$$

Proof. We start from the gamma reflection formula (2.32), evaluating the logarithm of both sides:

$$\ln\left(\Gamma(x)\,\Gamma(1-x)\right) = \ln\left(\frac{\pi}{\sin(\pi\,x)}\right).$$

Next, applying the properties of the logarithm, we obtain

$$\ln\Gamma(x) + \ln\Gamma(1-x) = \ln\pi - \ln\sin(\pi x).$$

Now, differentiating again, recalling the definition of digamma, we find out

$$\psi(x) - \psi(1-x) = -\pi \frac{\cos(\pi x)}{\sin(\pi x)} = -\pi\cot(\pi x).$$

By inductive argument, the thesis follows differentiating $n \in \mathbb{N}$ times the last equality. □

Chapter 4

Zeta and Hurwitz function

The zeta function is fundamental in analytical number theory and has applications in theoretical physics and probability. An exhaustive treatment would require (at least) an entire monograph. Still, due to the introductory character of these lecture notes, we limit ourselves to a brief overview, referring to [45, 66, 70, 109, 114] for further details. In any case, we present the fundamental Riemann result (Theorem 4.4) relating to the ζ functional equation, formula (4.15), which allows its analytical continuation, discussing briefly its implications with prime numbers theory, leading to the celebrated Riemann's hypothesis, [23]. It is worth noting that zeta is linked with gamma as stated in Theorem 4.3. The chapter ends with a brief introduction to the Hurwitz zeta function, which is closely related to zeta.

4.1 Introduction

Zeta function is defined, for $s > 1$, by the harmonic series:

$$\zeta(s) = \sum_{n=1}^{\infty} \frac{1}{n^s}\,. \tag{4.1}$$

Since zeta is defined as a series with positive terms dependent on parameter s, the condition $s > 1$ ensures its convergence. Convergence is also assured for all complex numbers $s = \sigma + i\,\tau$ such that $\sigma > 1$: in this case, we have

$$\zeta(s) = \zeta(\sigma + i\,\tau) = \sum_{n=1}^{\infty} \frac{1}{n^{\sigma}\,(\cos(\tau\,\ln n) + i\,\sin(\tau\,\ln n))}\,. \tag{4.1b}$$

The interest in this function dates back to 1730 when Euler, after having solved the Basel problem, which in terms of the given definition corresponds to the computation in closed form of $\zeta(2)$, arrived in 1749 at the determination of $\zeta(2n)$ with $n \in \mathbb{N}$. Euler also discovered the aspect that made and still makes this function the object of study by the mathematical community, highlighting its surprising connection to prime numbers. For this reason, zeta is extremely popular with number theorists, and its applications go as far

DOI: 10.1201/9781003171546-4

as quantum mechanics. Ayoub's article [17] is an essential reference to the study of Euler's contribution to this field. Subsequently, in 1859, Riemann (1826–1866) in a seminal paper [98] found an explicit formula to see how many prime numbers are smaller than a fixed integer, which represents an essential improvement over the approximate value known at that time. Riemann studied the zeta function for the first time from the point of view of complex analysis. However, the Riemann formula depends on the knowledge of a particular distribution of the non-trivial zeros of the continuation of zeta to $\mathbb{C} \setminus \{1\}$: this complex continuation has zeroes (called trivial) at negative even integers, and this, as we will see, depends on the link that Riemann established between zeta and gamma. Riemann hypothesised, without, however, proving it, that all the non-trivial zeros of zeta have a real part equal to $1/2$, and this problem is still unsolved and is known as "Riemann hypothesis". In the rest of our discussion, when we have explained the terms of the question, we will formalise the problem in rigorous terms. Due to the importance of this problem, many scholars and many authoritative sources refer to the zeta function, calling it Riemann's zeta, committing, however, a historical inaccuracy, indeed, because the scope of the Riemann conjecture has, as we have said, enormous importance and its name is inextricably linked to this function. Riemann died prematurely at forty, and unfortunately, most of his notes have been lost. However, the analysis of the remaining manuscripts showed that Riemann had developed very advanced methods for studying the position of zeros, which has generated significant interest in his work.

Remark. To calculate the zeta values for $s > 1$, we can limit to adding only considering the odd indices, as in the case already seen of $\zeta(2)$, by separating the sum between even and odd, we find

$$\begin{aligned}\zeta(s) &= \sum_{n=1}^{\infty} \frac{1}{(2n)^s} + \sum_{n=0}^{\infty} \frac{1}{(2n+1)^s} = \frac{1}{2^s} \sum_{n=1}^{\infty} \frac{1}{n^s} + \sum_{n=0}^{\infty} \frac{1}{(2n+1)^s} \\ &= 2^{-s}\zeta(s) + \sum_{n=0}^{\infty} \frac{1}{(2n+1)^s} .\end{aligned}$$

In particular, we obtain

$$\zeta(s) = \frac{2^s}{2^s - 1} \sum_{n=0}^{\infty} \frac{1}{(2n+1)^s} . \tag{4.2}$$

Zeta is associated with its alternating sign version, known as the Dirichlet eta function:

$$\eta(s) = \sum_{n=1}^{\infty} \frac{(-1)^{n-1}}{n^s} . \tag{4.3}$$

Notice that eta is defined for $s > 0$, therefore in a larger domain than that of zeta, but these functions are naturally connected; in fact, we have

Theorem. If $s > 1$, then

$$\zeta(s) = \frac{1}{1 - 2^{s-1}} \eta(s) . \tag{4.4}$$

Proof. Summing zeta and eta:

$$\begin{aligned} \zeta(s) &= \frac{1}{1^s} + \frac{1}{2^s} + \frac{1}{3^s} + \cdots + \frac{1}{n^s} + \ldots \\ \eta(s) &= \frac{1}{1^s} - \frac{1}{2^s} + \frac{1}{3^s} + \cdots + \frac{1}{n^s} + \ldots \end{aligned}$$

We observe that the terms of the even index cancel each other, leading to equality:

$$\zeta(s) + \eta(s) = 2 \sum_{n=0}^{\infty} \frac{1}{(2n+1)^s} .$$

We replace the series on the right-hand side using (4.1b), obtaining

$$\zeta(s) + \eta(s) = 2 \left(1 - 2^{-s}\right) \zeta(s).$$

Solving with respect to $\zeta(s)$, we obtain (4.4) and conclude the proof. □

Remark. The series defining eta converges for every $s > 0$; for this reason, formula (4.4) just demonstrated allows to extend the zeta domain to $s > 0$ with the exclusion of $s = 1$ where the coefficient of eta has a singularity. The nature of this singularity is characterised in the next Theorem 4.1.

We need to recall the notion of asymptotically equivalent functions.

Definition. If $f(x)$ and $g(x)$ are real functions defined in a neighbourhood of $a \in \mathbb{R}$ we say that $f(x)$ is asymptotically equivalent to $g(x)$ for $x \to a$ if:

$$\lim_{x \to a} \frac{f(x)}{g(x)} = 1.$$

In that case, we will write $f(x) \sim g(x)$.

Theorem 4.1. For $s \to 1^+$ we have that

$$\zeta(s) \sim \frac{1}{s-1},$$

or in other terms:

$$\lim_{s \to 1^+} \frac{\zeta(s)}{s-1} = 1.$$

Proof. Rewrite (4.4) as

$$\left(1 - \frac{2}{2^s}\right) \zeta(s) = \eta(s). \tag{4.4b}$$

But $\eta(1) = \ln 2$ so, since using de l'Hopital rule we have

$$\lim_{s\to 1} \frac{1-2^{s-1}}{s-1} = \ln 2.$$

So comparing with (4.4b), we arrive at:

$$\lim_{s\to 1^+} (s-1)\,\zeta(s) = \lim_{s\to 1^+} \frac{s-1}{1-2^{s-1}} \left(1-\frac{2}{2^s}\right)\zeta(s) = \frac{1}{\ln 2}\ln 2 = 1.$$

□

4.2 Euler contributes to zeta the function

Euler in 1737 discovered the connection between zeta and prime numbers, giving life to the branch of mathematics now known as analytic number theory.

Theorem. For any $s > 1$, if $\mathbb{P}$ stands for the set of prime numbers, then

$$\zeta(s) = \prod_{p\in\mathbb{P}} \left(1-\frac{1}{p^s}\right)^{-1}.$$

Proof. From (4.4b) we infer

$$\zeta(s)\left(1-\frac{1}{2^s}\right) = \eta(s) = 1 + \sum_{2\nmid n} \frac{1}{n^s} \tag{4.5}$$

we $2 \nmid n$ indicates that 2 does not divide $n \in \mathbb{N}$. Similarly by multiplying $1/3^s$ (4.5) we get

$$\zeta(s)\left(1-\frac{1}{2^s}\right)\frac{1}{3^s} = \eta(s) = \sum_{2\nmid n} \frac{1}{(3n)^s} \tag{4.5b}$$

and subtracting (4.5) and (4.5b)

$$\zeta(s)\left(1-\frac{1}{2^s}\right)\left(1-\frac{1}{3^s}\right) = 1 + \sum_{\substack{2\nmid n\\ 3\nmid n}} \frac{1}{n^s}.$$

Continuing in this way, we obtain

$$\zeta(s)\prod_{n=1}^{N}\left(1-\frac{1}{p_n^s}\right) = 1 + \sum_{p_1,\,p_2,\,\ldots,p_N\nmid n} \frac{1}{n^s} \tag{4.6}$$

where $p_1, \ldots, p_N$ are the first N prime numbers. But $p_1, \ldots p_N \nmid n$ and $n > 1$ implies $n > N$ and then the series on the right-hand side in (4.6) cannot exceed

$$\sum_{n=N+1}^{\infty} \frac{1}{n^s}.$$

Hence taking the limit for $N \to \infty$ in (4.6) we arrive at

$$\lim_{N\to\infty} \zeta(s) \prod_{n=1}^{N} \left(1 - \frac{1}{p_n^s}\right) = 1.$$

Our statement is proven. □

The problem that led Euler to study zeta was the Basel problem involving the explicit calculation of $\zeta(2)$. The first solution Euler found was based on his formula for the sine infinite product (1.41). Euler then extended his result to the case of $\zeta(2n)$ with $n \in \mathbb{N}$. His demonstration makes massive use of the so-called Bernoulli numbers, introduced by Jakob Bernoulli in his famous posthumous treatise Ars Conjectandi, to study sums of powers of the type $S_m(n) = 1^m + 2^m + \cdots + n^m$. Our discussion follows the setting and use of this sequence of numbers due to Euler, which we report in detail to benefit the reader: our presentation of this subject is inspired by [39, chapter 11].

Definition. We call Bernoulli numbers $(B_n)_{n\in\mathbb{N}}$ the sequence of numbers defined by

$$f(x) = \frac{x}{e^x - 1} = \sum_{n=0}^{\infty} B_n \frac{x^n}{n!}. \tag{4.7}$$

Remark. Sequence (B_n) individuates the coefficients of the exponential generating function of $f(x)$. Notice that the cognate sequence (b_n) identified by

$$\frac{x}{e^x - 1} = \sum_{n=0}^{\infty} b_n \, x^n,$$

individuates the coefficients of the ordinary generating function of $f(x)$. Observe that the singularity in the origin of $f(x)$ is apparent, since

$$\lim_{x\to 0} f(x) = \lim_{x\to 0} \frac{x}{e^x - 1} = 1.$$

Hence, $f(x)$ is analytic, meaning that (4.7) is well-posed.

We notice that (B_n) is not expressible in closed form. However, it is possible to detect (B_n) through recursive methods. Before doing so, however, we observe that the only odd Bernoulli number other than zero is B_1. In this regard, from the calculation of the first derivative of $f(x)$, we have

$$\lim_{x\to 0} f'(x) = \lim_{x\to 0} \frac{e^x(1-x) - 1}{(e^x - 1)^2} = \lim_{x\to 0} \frac{-\frac{1}{2}\,x^2 + \mathrm{o}(x^2)}{x^2 + \mathrm{o}(x^2)} = -\frac{1}{2}.$$

So right now we know that $B_0 = 1$ and $B_1 = -1/2$ and this allows us to deduce that $B_{2n+1} = 0$ for each $n \geq 1$, in fact observing that

$$\frac{x}{e^x - 1} + \frac{1}{2}x = 1 + \sum_{n=2}^{\infty} B_n \frac{x^n}{n!}, \tag{4.8}$$

we see that on the left-hand side of (4.8) appears an even function; it follows that its power series has no nonzero odd-power terms. We also observe that this last relation makes us realise that the sequence (B_n) is connected to the hyperbolic cotangent. This will allow us to exploit the fundamental series development presented in (1.35). This fact is crucial as it allows us to represent $\zeta(2n)$ values in terms of Bernoulli numbers for $n \in \mathbb{N}$.

Before demonstrating the recursive law that determines Bernoulli numbers, let us recall the notion of the Cauchy convolution product of two series, see [77, p. 316]:

$$\left(\sum_{n=0}^{\infty} a_n\right)\left(\sum_{n=0}^{\infty} b_n\right) = \sum_{n=0}^{\infty} c_n, \quad \text{where } c_n = \sum_{k=0}^{n} a_k b_{n-k} \,.$$

Theorem. Sequence (B_n) verifies $B_1 = 1$ and if $n \in \mathbb{N}$, $n \geq 2$ then

$$\sum_{k=0}^{n-1} \binom{n}{k} B_k = 0. \tag{4.9}$$

Proof. The proof starts from (4.7), multiplying both sides by $(e^x - 1)$, which is subsequently represented in a power series, and then applying the relation just recalled the Cauchy product of two series:

$$\begin{aligned} x = (e^x - 1)\sum_{n=0}^{\infty} B_n \frac{x^n}{n!} &= \sum_{j=1}^{\infty} \frac{x^j}{j!} \sum_{n=0}^{\infty} B_n \frac{x^n}{n!} \\ &= \sum_{n=0}^{\infty} \frac{x^{n+1}}{(n+1)!} \sum_{n=0}^{\infty} B_n \frac{x^n}{n!} \\ &= \sum_{n=0}^{\infty} \sum_{k=0}^{n} \frac{x^{n+1-k}}{(n+1-k)!} B_k \frac{x^k}{k!} . \end{aligned}$$

To arrive at the thesis, after multiplying the two powers of x, we multiply and divide the term of the sequence to be added in the last expression obtained by $(n+1)!$ so that we can represent it using the binomial coefficient

$$\begin{aligned} x &= \sum_{n=0}^{\infty} \sum_{k=0}^{n} \frac{(n+1)!}{(n+1-k)!\,k!} B_k \frac{x^{n+1}}{(n+1)!} \\ &= \sum_{n=0}^{\infty} \sum_{k=0}^{n} \binom{n+1}{k} B_k \frac{x^{n+1}}{(n+1)!} . \end{aligned}$$

Finally, we scale the summation index, obtaining:

$$x = \sum_{n=1}^{\infty} \sum_{k=0}^{n-1} \binom{n}{k} B_k \frac{x^n}{n!}$$

from which, because on the left side appears only x, we obtain (4.9). □

We go back to (4.8) observing that since the left-hand side can be written as

$$\frac{x}{2} \frac{(e^x + 1)}{(e^x - 1)} = \frac{x}{2} \coth \frac{x}{2}.$$

Recalling also that (B_n) has no nonzero odd term we infer

$$\frac{x}{2} \coth \frac{x}{2} = 1 + \sum_{n=2}^{\infty} B_n \frac{x^n}{n!} = 1 + \sum_{n=1}^{\infty} B_{2n} \frac{x^{2n}}{(2n)!}. \tag{4.10}$$

Placing $x = 2y$ in (4.10) we obtain the series representation of the hyperbolic cotangent:

$$y \coth y = 1 + \sum_{n=1}^{\infty} \frac{(2y)^{2n}}{(2n)!} B_{2n}. \tag{4.10b}$$

Finally, by setting $y = i\,x$ in (4.10b), using (1.16d) and (1.16e), we obtain a new series representation of the cotangent, which holds for $|x| < \pi$, $n \neq 0$

$$x \cot x = 1 + \sum_{n=1}^{\infty} (-1)^n \frac{(2x)^{2n}}{(2n)!} B_{2n}. \tag{4.11}$$

The last step, given the proof of Euler's sums Theorem, consists of replacing x with πx so that

$$\pi x \cot(\pi x) = 1 + \sum_{n=1}^{\infty} (-1)^n \frac{\pi^{2n} (2x)^{2n}}{(2n)!} B_{2n}. \tag{4.11b}$$

At this point, we can state and prove Euler's sums Theorem essentially comparing the two expressions for the cotangent we have: one obtained using the Fourier series, equation (1.34), and (4.11) involving Bernoulli numbers.

Theorem 4.2 (Euler's sums Theorem)**.** For any $n \in \mathbb{N}$, $\zeta(2n)$ is given by

$$\zeta(2n) = (-1)^{n+1}\, 2^{2n-1} \frac{\pi^{2n}}{(2n)!} B_{2n}. \tag{4.12}$$

Proof. We start by rewriting (1.34), page 35, as

$$\begin{aligned}
\pi x \cot(\pi x) &= 1 - 2 \sum_{n=1}^{\infty} \frac{x^2}{n^2 - x^2} = 1 - 2 \sum_{n=1}^{\infty} \frac{(x/n)^2}{1 - (x/n)^2} \\
&= 1 - 2 \sum_{n=1}^{\infty} \sum_{k=1}^{\infty} \left(\frac{x}{n}\right)^{2k} && \text{(a)} \\
&= 1 - 2 \sum_{k=1}^{\infty} \sum_{n=1}^{\infty} \left(\frac{x}{n}\right)^{2k} && \text{(b)} \\
&= 1 - 2 \sum_{k=1}^{\infty} \left(\sum_{n=1}^{\infty} \frac{1}{n^{2k}} \right) x^{2k}. && \text{(c)}
\end{aligned}$$

In (a), we used the geometric series; in (b), we exchanged the summation order, which is legitimate due to the uniform convergence of the two series; and in (c), we reorganised the expression to be summed to highlight the role of the zeta function: ultimately Fourier series (1.34) can be expressed in the form, after a change of index:

$$\pi x \cot(\pi x) = 1 - 2\sum_{n=1}^{\infty} \zeta(2n)\, x^{2n}. \tag{4.11c}$$

Now comparing (4.11c) and (4.11b) yields

$$-2\zeta(2n) = (-1)^n\, 2^{2n}\pi^{2n}\, \frac{B_{2n}}{(2n)!}.$$

In conclusion, (4.12) follows after simplifying. □

The exact calculation of $\zeta(2n)$ with $n \in \mathbb{N}$, therefore, depends on the values of (B_{2n}) obtainable from the formula (4.9) or by direct calculation of the McLaurin polynomial of $f(x)$ defined in (4.7); they are implemented in all computer algebra systems and therefore available without much difficulty. The behaviour of zeta for odd arguments is still unknown: the only significant result due to Roger Apéry (1916–1994) [8] is that $\zeta(3)$ is irrational, Apéry's proof relies on the particular representation of $\zeta(3)$:

$$\zeta(3) = \frac{5}{2}\sum_{n=1}^{\infty} \frac{(-1)^{n-1}}{n^3\binom{2n}{n}}.$$

Here are some values of $\zeta(2n)$:

$$B_2 = \frac{1}{6} \qquad B_6 = \frac{1}{42} \qquad \zeta(2) = \frac{\pi^2}{6} \qquad \zeta(6) = \frac{\pi^6}{945}$$

$$B_4 = -\frac{1}{30} \qquad B_8 = -\frac{1}{30} \qquad \zeta(4) = \frac{\pi^4}{90} \qquad \zeta(8) = \frac{\pi^8}{9450}.$$

4.3 Analytic continuation of zeta

There are various approaches to obtaining the analytic continuation of zeta; see definition 1.16; this is particularly important for its connections with the analytic number theory, which we will not deal with here: the interested reader can refer, for instance, to the classic treatise of Apostol [9] or to the more recent contribution of Borwein, Choi, Rooney, and Weirathmueller [23], even if at the end of the chapter we will make a quick mention of Riemann's hypothesis.

To clarify the first representation of the zeta continuation, let us recall the notion of integer part $\lfloor x \rfloor$ of $x \in \mathbb{R}$, i.e. the greatest integer non-exceeding x :

$$\lfloor x \rfloor := \min \{y \in \mathbb{Z} : \, y \le x\}.$$

Observe that $\lfloor x \rfloor \le x < \lfloor x \rfloor + 1$ and $\lfloor x + n \rfloor = n + \lfloor x \rfloor$ for any $n \in \mathbb{Z}$. The fractional part of x is $\{x\} = x - \lfloor x \rfloor$. Thus $0 \le \{x\} < 1$. Notice also that function $x \mapsto \{x\}$ is periodic with period 1. Integer part and ceiling function introduced at page 121 are naturally related:

$$\lceil x \rceil - \lfloor x \rfloor = \begin{cases} 0 & \text{if } x \in \mathbb{Z}, \\ n & \text{if } x \notin \mathbb{Z}. \end{cases}$$

Theorem. If $s > 1$ then

$$\zeta(s) = \frac{s}{s-1} - s \int_1^\infty \frac{\{x\}}{x^{s+1}} \, \mathrm{d}x \,. \tag{4.13}$$

Proof. We begin observing that if $s > 1$ and $x \ge 1$ then

$$0 \le \frac{\{x\}}{x^{s+1}} \le \frac{1}{x^2}.$$

In this way, the convergence of the integral to the right side of (4.13) is verified; moreover, given the periodicity of the fractional part, we can evaluate such integral decomposing the real line in intervals with integer endpoints:

$$\begin{aligned} \int_1^\infty \frac{\{x\}}{x^{s+1}} \, \mathrm{d}x &= \sum_{n=1}^\infty \int_n^{n+1} \frac{\{x\}}{x^{s+1}} \, \mathrm{d}x = \sum_{n=1}^\infty \int_n^{n+1} \frac{x - \lfloor x \rfloor}{x^{s+1}} \, \mathrm{d}x \\ &= \sum_{n=1}^\infty \left(\int_n^{n+1} \frac{1}{x^s} \mathrm{d}x - \int_n^{n+1} \frac{\lfloor x \rfloor}{x^{s+1}} \mathrm{d}x \right) \\ &= \sum_{n=1}^\infty \left(\int_n^{n+1} \frac{1}{x^s} \mathrm{d}x - n \int_n^{n+1} \frac{1}{x^{s+1}} \mathrm{d}x \right) \\ &= \sum_{n=1}^\infty \left(\frac{1}{1-s} \left(\frac{1}{(n+1)^{s-1}} - \frac{1}{n^{s-1}} \right) - \frac{n}{s} \left(\frac{1}{n^s} - \frac{1}{(n+1)^s} \right) \right). \end{aligned}$$

Observe that

$$\frac{1}{1-s} \sum_{n=1}^\infty \left(\frac{1}{(n+1)^{s-1}} - \frac{1}{n^{s-1}} \right) = \frac{1}{s-1},$$

and

$$\frac{1}{s} \sum_{n=1}^\infty n \left(\frac{1}{n^s} - \frac{1}{(n+1)^s} \right) = \frac{1}{s} \left(\sum_{n=1}^\infty \frac{1}{n^{s-1}} - \sum_{n=1}^\infty \frac{n}{(n+1)^s} \right). \tag{a}$$

Rescaling the index in the second series in (a), putting $m = n + 1$, we get

$$\sum_{n=1}^\infty \frac{n}{(n+1)^s} = \sum_{m=2}^\infty \frac{m-1}{m^s} = \sum_{m=2}^\infty \frac{1}{m^{s-1}} - \sum_{m=2}^\infty \frac{1}{m^s}$$

hence using again n as the summation index and going back to (a)

$$\frac{1}{s}\sum_{n=1}^{\infty} n\left(\frac{1}{n^s}-\frac{1}{(n+1)^s}\right)=\frac{1}{s}\left(\sum_{n=1}^{\infty}\frac{1}{n^{s-1}}-\sum_{n=2}^{\infty}\frac{1}{n^{s-1}}+\sum_{n=2}^{\infty}\frac{1}{n^s}\right)$$
$$=\frac{1}{s}\left(1+\zeta(s)-1\right)=\frac{1}{s}\zeta(s)\,.$$

Ultimately:

$$\int_1^{\infty}\frac{\{x\}}{x^{s+1}}\,\mathrm{d}x=\sum_{n=1}^{\infty}\left(\frac{1}{1-s}\left(\frac{1}{(n+1)^{s-1}}-\frac{1}{n^{s-1}}\right)-\frac{n}{s}\left(\frac{1}{n^s}-\frac{1}{(n+1)^s}\right)\right)$$
$$=\frac{1}{s-1}-\frac{1}{s}\zeta(s).$$

Eventually (4.13) follows solving for $\zeta(s)$. □

Remark. Looking at formula (4.13), we see that the function:

$$s\mapsto\int_1^{\infty}\frac{\{x\}}{x^{s+1}}\,\mathrm{d}x,$$

is analytic in the half-plane $\operatorname{Re} s>0$; thus, formula (4.13) extends the domain of zeta to this region.

A further way to prolong zeta is obtained by studying the connection of zeta (and eta) to gamma through integral representations. The following remarkable result holds.

Theorem 4.3.

$$\zeta(s)=\frac{1}{\Gamma(s)}\int_0^{\infty}\frac{x^{s-1}}{e^x-1}\mathrm{d}x \tag{4.14a}$$

$$\eta(s)=\frac{1}{\Gamma(s)}\int_0^{\infty}\frac{x^{s-1}}{e^x+1}\mathrm{d}x \tag{4.14b}$$

Proof. We consider the integral:

$$\int_0^{\infty}\frac{x^{s-1}}{e^x-1}\mathrm{d}x=\int_0^{\infty}\frac{x^{s-1}e^{-x}}{1-e^{-x}}\mathrm{d}x.$$

Then, using the geometric series starting from y, $|y|<1$:

$$\frac{y}{1-y}=\sum_{n=1}^{\infty}y^n,$$

we find out

$$\int_0^{\infty}\frac{x^{s-1}e^{-x}}{1-e^{-x}}\mathrm{d}x=\int_0^{\infty}\sum_{n=1}^{\infty}x^{s-1}e^{-nx}\mathrm{d}x.$$

Here, we can exchange the series with the integral:

$$\int_0^\infty \sum_{n=1}^\infty x^{s-1}e^{-nx}\mathrm{d}x = \sum_{n=1}^\infty \int_0^\infty x^{s-1}e^{-nx}\mathrm{d}x.$$

We change the variable by setting $t = nx$ so that

$$\sum_{n=1}^\infty \int_0^\infty x^{s-1}e^{-nx}\mathrm{d}x = \sum_{n=1}^\infty \int_0^\infty \left(\frac{t}{n}\right)^{s-1} e^{-t}\frac{1}{n}\mathrm{d}t$$
$$= \sum_{n=1}^\infty \frac{1}{n^s}\int_0^\infty t^{s-1}e^{-t}\mathrm{d}t = \zeta(s)\Gamma(s).$$

Therefore, we have shown (4.14a). The proof of the formula for the integral representation (4.14b) of eta is quite similar; it starts from the integral:

$$\int_0^\infty \frac{x^{s-1}}{e^x+1}\mathrm{d}x,$$

with the foresight to transform the fraction, which appears in the integrand, containing the exponential, to use the formula for the sum of the geometric series:

$$\int_0^\infty \frac{x^{s-1}}{e^x+1}\mathrm{d}x = \int_0^\infty \frac{e^{-x}}{1+e^{-x}}x^{s-1}\mathrm{d}x = \sum_{n=0}^\infty (-1)^n \int_0^\infty x^{s-1}e^{-(n+1)x}\mathrm{d}x$$
$$= \sum_{n=0}^\infty (-1)^n \frac{1}{(n+1)^s}\,\Gamma(s) = \Gamma(s)\sum_{n=1}^\infty (-1)^{n-1}\frac{1}{n^s}$$
$$= \Gamma(s)\,\eta(s).$$

□

4.4 Riemann contribute to zeta function

What is universally recognised as the fundamental contribution given by Riemann in the study of zeta is the so-called functional equation, through which the zeta can be extended to the complex plane to the exclusion of the pole $s = 1$. What we report here is substantially the original contribution of Riemann, [98], which has been practically taken from all the manuals on the subject, we quote without pretending to be exhaustive, [39, 45, 54, 85, 114]. It is a demonstration of central interest in the theory of special functions because it is based both on the properties of the Jacobi Theta function, and therefore on the elliptic functions, and the main properties of Gamma: Legendre's duplication and reflection by Euler. Therefore, in addition to the importance of the functional equation itself, it is also one of the most elegant and formative examples of proof in Mathematical Analysis.

Theorem 4.4 (Riemann's functional equation)**.** The zeta function fulfills the equation:

$$\zeta(s) = 2^s \pi^{s-1} \sin\left(\tfrac{s}{2}\,\pi\right) \Gamma(1-s)\,\zeta(1-s). \tag{4.15}$$

Remark. Equation (4.15) is known as Riemann's functional equation, writing $s = \sigma + i\,\tau$ we see that it allows continuing zeta analytically for $\sigma < 1$: since when $\sigma > 1$ we can use the initial definition of zeta, equation (4.1), while if $\sigma < 1$ the right side of (4.15) is well defined again on terms of (4.1). Given Theorem 4.1, we thus infer that zeta is analytic everywhere except for $s = 1$, a simple pole with residue one as shown in Theorem 4.1.

Proof. Recalling proposition 2.2, equation (2.7) on page 61, we see that for $s > 0$ and $n \in \mathbb{N}$ the following identity holds:

$$\int_0^\infty x^{\frac{1}{2}\,s-1} e^{-n^2 \pi x} \mathrm{d}x = \frac{\Gamma(\frac{s}{2})}{\pi^{\frac{s}{2}}\, n^s}. \tag{4.16}$$

Summing for $n \in \mathbb{N}$ both sides of (4.16) after the exchange between the series and the integral on the left side, we find

$$\int_0^\infty x^{\frac{1}{2}\,s-1} \sum_{n=1}^{\infty} e^{-n^2 \pi x} \mathrm{d}x = \frac{\Gamma(\frac{s}{2})}{\pi^{\frac{s}{2}}}\,\zeta(s). \tag{4.17}$$

Now, in the integrand on the left-hand side of (4.17), we recognise the Jacobi capital Ψ function introduced in Section 1.10 equation (1.56). For our argument functional property (1.57) is essential, we start rewriting (4.17) using Ψ:

$$\begin{aligned}
\frac{\Gamma(\frac{s}{2})}{\pi^{\frac{s}{2}}}\,\zeta(s) &= \int_0^\infty x^{\frac{1}{2}\,s-1} \Psi(x)\,\mathrm{d}x \\
&= \int_0^1 x^{\frac{1}{2}\,s-1} \Psi(x)\,\mathrm{d}x + \int_1^\infty x^{\frac{1}{2}\,s-1} \Psi(x)\,\mathrm{d}x && \text{(a)} \\
&= \int_0^1 x^{\frac{1}{2}\,s-1} \left(\tfrac{1}{\sqrt{x}} \Psi\left(\tfrac{1}{x}\right) + \tfrac{1}{2\sqrt{x}} - \tfrac{1}{2}\right) \mathrm{d}x + \int_1^\infty x^{\frac{1}{2}\,s-1} \Psi(x)\,\mathrm{d}x && \text{(b)} \\
&= \tfrac{1}{2} \int_0^1 \left(x^{\frac{s}{2}-\frac{3}{2}} - x^{\frac{s}{2}-1}\right) \mathrm{d}x + \int_0^1 x^{\frac{s}{2}-\frac{3}{2}} \Psi\left(\tfrac{1}{x}\right) \mathrm{d}x \\
&\quad + \int_1^\infty x^{\frac{1}{2}\,s-1} \Psi(x)\,\mathrm{d}x \\
&= \frac{1}{s(s-1)} + \int_1^\infty x^{-\frac{s}{2}-\frac{1}{2}} \Psi(x)\,\mathrm{d}x + \int_1^\infty x^{\frac{1}{2}\,s-1} \Psi(x)\,\mathrm{d}x. && \text{(c)}
\end{aligned}$$

In (a), we have broken up the integration regions; in (b), we have exploited the functional property of Ψ in the interval $[0, 1]$ in (c), we have calculated the first integral with elementary methods and the changed variable in the second, also if we continue to use the variable x hence we can rewrite (4.17) in the following way:

$$\frac{\Gamma(\frac{s}{2})}{\pi^{\frac{s}{2}}}\,\zeta(s) = \frac{1}{s(s-1)} + \int_1^\infty \left(x^{\frac{1}{2}\,s-1} + x^{-\frac{s}{2}-\frac{1}{2}}\right)\Psi(x)\,\mathrm{d}x. \tag{4.18}$$

The function on the right side of (4.18) is invariant, changing s to $1-s$, which is easy to verify with a simple calculation, and this implies that the left-side must also be invariant concerning passing from s to $1-s$, so we have proved the identity:

$$\pi^{-\frac{s}{2}}\,\Gamma(\tfrac{s}{2})\,\zeta(s) = \pi^{-\frac{1-s}{2}}\,\Gamma(\tfrac{1-s}{2})\,\zeta(1-s). \tag{4.19}$$

Solving for $\zeta(s)$:

$$\zeta(s) = \pi^{s-\frac{1}{2}}\,\frac{\Gamma(\frac{1-s}{2})}{\Gamma(\frac{s}{2})}\,\zeta(1-s). \tag{4.20}$$

Recall Legendre duplication formula (2.50) taking $x = 1 - \frac{s}{2}$:

$$\pi^{\frac{1}{2}}\,2^s\,\Gamma(1-s) = \Gamma\left(\tfrac{1-s}{2}\right)\Gamma\left(1-\tfrac{s}{2}\right).$$

Solve for $\Gamma\left(\frac{1-s}{2}\right)$ then insert in (4.20) so that

$$\zeta(s) = \frac{(2\pi)^s\Gamma(1-s)\zeta(1-s)}{\Gamma\left(1-\frac{s}{2}\right)\Gamma\left(\frac{s}{2}\right)}. \tag{4.21}$$

Finally reflexion formula (2.32) yields:

$$\Gamma\left(1-\tfrac{s}{2}\right)\Gamma\left(\tfrac{s}{2}\right) = \frac{\pi}{\sin\left(\frac{s}{2}\,\pi\right)}.$$

Thus, inserting in (4.21), we arrive at (4.15). □

Riemann functional equation (4.15) seems not defined for $s = 0$: indeed, it is an indeterminate form, which in this formulation, does not allow to compute the value of $\zeta(0)$. This fact seems to contradict our statement that the formula (4.15) will enable us to continue zeta analytically: singularity in $s = 0$ can be removed simply replacing $\Gamma(1-s)$ using the reflection formula (2.32), that is:

$$\frac{\pi}{\sin(\pi\,s)\;\Gamma(s)}.$$

Again invoking (2.32) and finally solving for $\zeta(1-s)$ leading to

$$\zeta(1-s) = 2^{1-s}\pi^{-s}\,\Gamma(s)\zeta(s)\cos\left(\tfrac{\pi s}{2}\right). \tag{4.22}$$

Hence $\zeta(0)$ is obtainable from (4.22) taking the limit for $s \to 1$ in the right hand side, also recalling Theorem 4.1, in fact

$$\begin{aligned}
\zeta(0) &= \lim_{s\to 1}\zeta(1-s) = \lim_{s\to 1} 2^{1-s}\pi^{-s}\,\Gamma(s)\zeta(s)\cos\left(\tfrac{\pi s}{2}\right)\\
&= \lim_{s\to 1} 2^{1-s}\pi^{-s}\,\Gamma(s)\ \lim_{s\to 1}\zeta(s)\cos\left(\tfrac{\pi s}{2}\right)\\
&= \frac{1}{\pi}\lim_{s\to 1}\zeta(s)\cos\left(\tfrac{\pi s}{2}\right)\\
&= \frac{1}{\pi}\lim_{s\to 1}\frac{\cos\left(\frac{\pi s}{2}\right)}{s-1} && \text{(Theorem 4.1)}\\
&= \frac{1}{\pi}\lim_{s\to 1}\left(-\frac{1}{2}\pi\sin\left(\frac{\pi s}{2}\right)\right) && \text{(de l'Hopital)}\\
&= -\frac{1}{2}.
\end{aligned}$$

The value found seems to hurt common sense if one makes the mistake of considering the definition of zeta in terms of series depending on the parameter s since, for this value of the parameter, the series diverges positively:

$$\zeta(0) = \frac{1}{1^0} + \frac{1}{2^0} + \frac{1}{3^0} + \cdots$$

The analytic continuation of a holomorphic function is the unique analytic function that, when restricted to the original domain, reduces to the given function, see page 17; therefore, in the extension, its behaviour can, as in this case, be very different from that of the prolonged function. Another interesting zeta value at continuation relates to $s = -1$. This value does not present particular computational difficulties; from (4.15), we easily see that

$$\zeta(-1) = 2^{-1}\pi^{-1-1}\sin\left(\tfrac{-1}{2}\,\pi\right)\,\Gamma(1-(-1))\,\zeta(1-(-1)) = -\frac{1}{2\pi^2}\frac{\pi^2}{6} = -\frac{1}{12}\,.$$

Again, if we use the original zeta expression, this relation leads to a paradoxical relationship:

$$\sum_{n=1}^{\infty} n = -\frac{1}{12},$$

since the left-hand side diverges positively.

We end this section with further observations on the functional property and a quick reference to the famous Riemann hypothesis. Rewriting formula (4.18) by explicitly deriving $\zeta(s)$, we find out

$$\zeta(s) = \frac{\pi^{\frac{s}{2}}}{\Gamma(\frac{s}{2})}\left(\frac{1}{s(s-1)} + \int_1^{\infty}\left(x^{\frac{1}{2}\,s-1} + x^{-\frac{s}{2}-\frac{1}{2}}\right)\Psi(x)\,\mathrm{d}x\right) \tag{4.18b}$$

As we already pointed out, in this way, we see that zeta is a meromorphic function with a simple pole at $s = 1$ and residue 1. For reasons of fundamental interest in analytic number theory, it is of interest to determine the zeros of zeta in $\mathbb{C} \setminus \{1\}$. This is a fascinating and still, in some respects, unresolved subject, which requires extensive treatment, for which we refer, for example, to [23, 9]. It is helpful to go back to equation (4.19) and introduce, to facilitate the discussion, the function:

$$\xi(s) = \pi^{-\frac{s}{2}} \frac{s(s-1)}{2} \Gamma(\tfrac{s}{2})\, \zeta(s). \tag{4.19b}$$

Theorem 4.4 reads in terms of $\xi(s)$ in a very simple and naive fashion, i.e.:

$$\xi(s) = \xi(1-s) \tag{4.15b}$$

From equation (4.15b), we infer that $\xi(s)$ is symmetric around the vertical line $\operatorname{Re} s = \frac{1}{2}$ on the other side, looking for zeros of zeta looking at (4.18b) we acknowledge that $\zeta(-2n) = 0$ for any $n \in \mathbb{N}$: this family of zeros is generated by the poles of the gamma function.

Riemann's hypothesis concerns the solutions of the equation $\zeta(s) = 0$, namely, Riemann hypothesised that in addition to the so-called trivial solutions given by $s = -2n$ with $n \in \mathbb{N}$, there are infinite solutions such that their real part is $1/2$. This conjecture has yet to be proven. However, numerical evidence suggests that this property holds. The importance of the Riemann conjecture lies in the fact that it is useful for studying the distribution of prime numbers. We do not go further on this topic, which does not strictly concern the special functions; we limit ourselves to suggesting the reading of [23] and the bibliographic information available therein.

4.5 Hurwitz zeta function

A. Hurwitz (1859-1919) introduced a parameter to generalise the zeta function, defining the function $\zeta(s, x)$ as:

$$\zeta(s, a) := \sum_{k=0}^{\infty} \frac{1}{(k+a)^s} \, . \tag{4.23}$$

Here $\operatorname{Re} s > 1$ and $\operatorname{Re} a > 0$. Hurwitz zeta function is related to zeta since, $\zeta(s) = \zeta(s, 1)$. Moreover, it fulfils the functional equation:

$$\zeta(s, a+1) = \zeta(s, a) - \frac{1}{a^s} \, . \tag{4.24}$$

Remark. Hurwitz function is linked to the polygamma functions if its first variable is a positive integer, i.e. $n \in \mathbb{N}$:

$$\psi_n(x) := \frac{d^n}{dx^n}\psi(x) = (-1)^{n+1} n!\, \zeta(n+1, x). \tag{4.25}$$

Recalling the formula (3.43) obtained on page 122, the thesis follows by comparing (4.23) and (3.43).

Remark. Let $s \neq 1$ and $a \in \mathbb{R}^+$, $b \in \mathbb{R}$ be such that $a + b \in \mathbb{R}^+$. Then

$$\int_0^1 \zeta(s+1, a+bx)\,\mathrm{d}x = \frac{1}{sb}\left\{\zeta(s,a) - \zeta(s, a+b)\right\}. \tag{4.26}$$

Formula (4.26) follows by

$$\int \zeta(s,x)\mathrm{d}x = \frac{1}{1-s}\zeta(s-1, x).$$

Which, in turn, follows from the relationship:

$$\frac{\partial \zeta}{\partial x}(s-1, x) = (1-s)\zeta(s,x),$$

which is a consequence of (4.23).

Remark. For $b > 0$, $\mathrm{Re}(s) > 1$, $\mathrm{Re}(x) > 0$ in consequence of the definition of $\zeta(s,x)$ we also have

$$\sum_{n=0}^{\infty} \frac{1}{(bn+x)^s} = \frac{1}{b^s}\,\zeta(s, \frac{x}{b})$$

The Hurwitz function admits various integral representations, which allow its continuation.

Theorem 4.5. The following integral representations hold for $\mathrm{Re}\, a > 0$ and $\mathrm{Re}\, s > 1$

$$\zeta(s,a) = \frac{1}{\Gamma(s)}\int_0^\infty \frac{t^{s-1}e^{-(a-1)t}}{e^t - 1}\mathrm{d}t, \tag{4.27a}$$

$$\zeta(s,a) = \frac{1}{2a^s} + \frac{1}{(s-1)a^{s-1}} + \frac{1}{\Gamma(s)}\int_0^\infty t^{s-1}e^{-at}\left(\frac{1}{e^t-1} - \frac{1}{t} + \frac{1}{2}\right)dt \tag{4.27b}$$

$$\zeta(s,a) = \frac{1}{2a^s} + \frac{1}{(s-1)a^{s-1}} + 2\int_0^\infty \frac{\sin(s\arctan\frac{t}{a})}{(a^2+t^2)^{s/2}(e^{2\pi t}-1)}\mathrm{d}t. \tag{4.27c}$$

The formula (4.27c) is due to Hermite.

Proof. For (4.27a), we follow the same method of Theorem 4.3, equations (4.14a) and (4.14b). Let's start with the integral:

$$\int_0^\infty \frac{t^{s-1}e^{-(a-1)t}}{e^t - 1}\mathrm{d}t = \int_0^\infty t^{s-1}e^{-(a-1)t}\frac{e^{-t}}{1-e^{-t}}\mathrm{d}t\,.$$

Then, using the geometric series method one more time, we find out, observing that as in Theorem 4.3, the exchange between integral and series is legitimate; furthermore, the computation of the integral in terms of gamma is the same as presented in the proof of the Theorem 4.3:

$$\begin{aligned}
\int_0^\infty t^{s-1}e^{-(a-1)t}\frac{e^{-t}}{1-e^{-t}}\mathrm{d}t &= \int_0^\infty \sum_{n=1}^\infty t^{s-1}e^{-(n+a-1)t}dt \\
&= \sum_{n=1}^\infty \int_0^\infty t^{s-1}e^{-(n+a-1)t}dt \\
&= \sum_{n=1}^\infty \frac{\Gamma(s)}{(n+a-1)^s} \\
&= \sum_{n=0}^\infty \frac{\Gamma(s)}{(n+a)^s} = \Gamma(s)\,\zeta(s,a).
\end{aligned}$$

Hence, (4.27a) is proved.

To prove (4.27b), we will make a couple of algebraic adjustments by adding and subtracting at (4.27a) the same quantities and reassembling the expressions obtained. We start by replacing t^{s-1} with $t^{s-1}(1+e^{-t}-e^{-t})$:

$$\begin{aligned}
\zeta(s,a) &= \frac{1}{\Gamma(s)}\int_0^\infty \frac{t^{s-1}e^{-(a-1)t}}{e^t-1}\mathrm{d}t \\
&= \frac{1}{\Gamma(s)}\int_0^\infty \frac{t^{s-1}(1+e^{-t}-e^{-t})e^{-(a-1)t}}{e^t-1}\mathrm{d}t \\
&= \frac{1}{\Gamma(s)}\int_0^\infty \frac{t^{s-1}e^{-(a-1)t}}{e^t-1}\mathrm{d}t + \frac{1}{\Gamma(s)}\int_0^\infty \frac{t^{s-1}e^{-t}e^{-(a-1)t}}{e^t-1}\mathrm{d}t \\
&\quad - \frac{1}{\Gamma(s)}\int_0^\infty \frac{t^{s-1}e^{-t}e^{-(a-1)t}}{e^t-1} \\
&= I_1 + I_2 - I_3.
\end{aligned}$$

Without too much difficulty, we see that $I_1 = \zeta(s,a)$, $\quad I_3 = \zeta(s,a+1)$, so that given (4.24)

$$I_1 - I_3 = \frac{1}{a^s}.$$

Vice versa, for I_2, the calculation is considerably more demanding. We rewrite I_2 once again by adding and removing the same expression, chosen to bring up the integrand of the formula (4.27b) we want to demonstrate; at the beginning, we adjust the expression of the integrand:

$$\begin{aligned} I_2 &= \frac{1}{\Gamma(s)} \int_0^\infty t^{s-1} e^{-at} \left(e^t - 1\right)^{-1} \mathrm{d}t \\ &= \frac{1}{\Gamma(s)} \int_0^\infty t^{s-1} e^{-at} \left[\left(e^t - 1\right)^{-1} - \frac{1}{t} + \frac{1}{2} - \frac{1}{2} + \frac{1}{t}\right] \mathrm{d}t \\ &= \frac{1}{\Gamma(s)} \int_0^\infty t^{s-1} e^{-at} \left[\left(e^t - 1\right)^{-1} - \frac{1}{t} + \frac{1}{2}\right] \mathrm{d}t \\ &- \frac{1}{2}\frac{1}{\Gamma(s)} \int_0^\infty t^{s-1} e^{-at} \mathrm{d}t + \frac{1}{\Gamma(s)} \int_0^\infty t^{s-2} e^{-at} \mathrm{d}t\,. \end{aligned}$$

The first addend, as we wish, is the integral that appears in the formula (4.27b), the second and third are calculated in terms of gamma according to Proposition 2.2, formula (2.7), page 61, leading to

$$I_2 = \frac{1}{\Gamma(s)} \int_0^\infty t^{s-1} e^{-at} \left[\left(e^t - 1\right)^{-1} - \frac{1}{t} + \frac{1}{2}\right] \mathrm{d}t - \frac{1}{2a^s} + \frac{1}{(s-1)a^{s-1}}.$$

Formula (4.27b) eventually follows computing $I_1 + I_2 - I_3$.

We skip the proof of the Hermite representation formula (4.27c), referring to the argument exposed in [35]. □

Remark. Form (4.27c), we see that the Hurwitz zeta function can be continued, exactly like zeta, for all complex numbers s with $s \neq 1$. At $s = 1$, it has a simple pole with residue 1. Moreover, taking $a = 1$ (4.27c) provides a further meromorphic continuation for zeta, namely:

$$\zeta(s) = \frac{1}{2} + \frac{1}{s-1} + 2\int_0^\infty \frac{\sin(s \arctan t)}{(1+t^2)^{s/2}(e^{2\pi t} - 1)} \mathrm{d}t.$$

Corollary 4.6. If $a > 0$ then

$$\lim_{s\to 1} \left(\zeta(s,a) - \frac{1}{s-1}\right) = -\psi(a) \tag{4.28}$$

In particular

$$\lim_{s\to 1} \left(\zeta(s) - \frac{1}{s-1}\right) = \gamma \tag{4.28b}$$

where γ is the Euler-Mascheroni constant.

Proof. Recalling (4.27c) we can take the limit as $s \to 1$ and obtain

$$\begin{aligned} &\lim_{s\to 1} \left(\zeta(s,a) - \frac{1}{s-1}\right) = \\ &\frac{1}{2a^2} + 2\int_0^\infty \frac{\sin(\arctan \frac{t}{a})}{(a^2+t^2)^{1/2}} \frac{\mathrm{d}t}{e^{2\pi t} - 1} + \lim_{s\to 1} \frac{1}{s-1}\left(\frac{1}{a^{s-1}} - 1\right) \end{aligned} \tag{4.29}$$

Now, using elementary calculus methods, we see that

$$\sin(\arctan \tfrac{t}{a}) = \frac{t}{\sqrt{a^2+t^2}}, \qquad \lim_{s\to 1} \frac{1}{s-1}\left(\frac{1}{a^{s-1}} - 1\right) = -\ln a.$$

Hence, plugging these two equalities in (4.29), we arrive at

$$\lim_{s\to 1}\left(\zeta(s,a) - \frac{1}{s-1}\right) = \frac{1}{2a^2} - \ln a + 2\int_0^\infty \frac{t}{a^2+t^2}\frac{dt}{e^{2\pi t}-1}.$$

The last is exactly the formula (3.38) of corollary 3.10 page 120. Finally (4.28b) follows after taking $a = 1$ from $\psi(1) = -\gamma$ see formula (3.5). □

Chapter 5

A synopsis on second-order linear ODE

The following chapters will focus on special functions obtained by solving second-order linear differential equations. We will explore the nature and properties of functions identified by linear differential equations, with particular reference to the case of variable coefficients. The interest in functions generated in this way naturally comes from motivations of application originating from physical-mathematical or statistical-mathematical problems. We assume the reader has basic training in ordinary differential equations, as exposed in the classic reference texts [10, 31, 34, 36, 37, 95, 97]. Still, it helps us understand what we will explain by fixing the notations, results, and techniques we will use in the following. Issues such as the reduction to the normal form of linear differential equations of the second order and the change of the independent variable only sometimes find space in the institutional syllabus on odes. The problem we will address here is related to explicitly determining the solutions of such equations with variable coefficients. It is commonplace that such equations admit solutions that cannot be represented in terms of elementary functions. In what follows, seeking a power series representation for the solutions of a given linear ordinary differential equation, we will work with analytic functions, i.e. functions $f : I \to \mathbb{R}$, $f \in \mathcal{C}^\infty$ being I a real open interval such that for $x_0 \in I$ the Taylor series (1.11) converges in some neighbourhoods of x_0, see Theorem 1.12. Most elementary functions like $\sin x$, $\cos x$ and e^x are analytic everywhere. We will report the essential elements without dwelling on the proofs and referring to the manuals on the subject. Here we suggest, among the vast literature, [95, chapter 5], [10, chapter 6], and [101, chapter 6]. In particular, before starting our discussion, we need to recall the techniques founded on the power series representation of the solutions in broad terms.

5.1 Solution set of a second order linear equation

Given two functions p and q, we can, at least initially, assume analytic in a certain interval, we are interested in the solution of the differential equation

$$Ly := y'' + p(x)\,y' + q(x)\,y = 0. \tag{5.1}$$

DOI: 10.1201/9781003171546-5

The solution set of (5.1) consists of a two-dimensional vector space. Recall that, given two functions $y_1(x)$ and $y_2(x)$ defined on some interval I, their Wronskian is defined as:

$$W(x) = \det \begin{bmatrix} y_1(x) & y_2(x) \\ y_1'(x) & y_2'(x) \end{bmatrix} .$$

The correct notation should be $W(y_1, y_2)(x)$ instead of $W(x)$ that is used for brevity and when a couple of functions y_1 and y_2 are individuated without ambiguity. The Wronskian is useful, since two solutions y_1 and y_2 to (5.1) are independent, if and only if $W(x) \neq 0$ for $x \in I$. As a matter of fact, it is indeed possible to show that in this case, the Wronskian solves the first-order separable differential equation $W' = -p(x)\,W$, and therefore

$$W(x) = W(x_0)\, e^{-P(x)}. \tag{5.2}$$

As in the following, we capitalise to represent the antiderivative, i.e. $P'(x) = p(x)$. Equation (5.2) implies that, if the Wronskian $W(x)$ vanishes in $x_0 \in I$, then $W(x)$ is the zero function; vice versa, if there exists $x_0 \in I$ such that $W(x_0) \neq 0$, then $W(x) \neq 0$ for each $x \in I$. If a solution $y_1(x)$ of (5.1) is known, a second independent solution is given by

$$y_2(x) = y_1(x) \int \frac{e^{-P(x)}}{y_1^2(x)}\, \mathrm{d}x\,. \tag{5.3}$$

5.2 Normal form

The change of independent variable $y = u\, e^{-\frac{1}{2}P(x)}$ transforms equation (5.1) into an equation of the form $u'' + I(x)\, u = 0$, where

$$I(x) = q(x) - \frac{1}{2}\, p'(x) - \frac{1}{4}\, p^2(x)\,.$$

This transformation will be extremely useful in chapter 7 when dealing with the zeros of Bessel functions. If, for instance, we consider the constant–coefficient equation, $a,\, b,\, c \in \mathbb{R}$,

$$a\, y'' + b\, y' + c\, y = 0, \tag{5.4}$$

then the change of variable $y = u\, e^{-\frac{b}{2a}x}$ leads to

$$u'' - \frac{b^2 - 4\,a\,c}{4\,a^2}\, u = 0. \tag{5.5}$$

The normal form (5.5) explains the nature of the following formulas (5.6), (5.7), and (5.8), describing the structure of the solution to a constant-coefficient homogeneous linear differential equation of second order. In the following, the discriminant is $\Delta = b^2 - 4\,a\,c$ and c_1, c_2 are constant:

1. if $\Delta > 0$,

$$y(x) = e^{-\frac{b}{2a}x} \left(c_1 \cosh\left(\frac{\sqrt{\Delta}}{2a}\, x\right) + c_2 \sinh\left(\frac{\sqrt{\Delta}}{2a}\, x\right) \right); \tag{5.6}$$

2. if $\Delta < 0$,

$$y(x) = e^{-\frac{b}{2a}x} \left(c_1 \cos\left(\frac{\sqrt{-\Delta}}{2a}\, x\right) + c_2 \sin\left(\frac{\sqrt{-\Delta}}{2a}\, x\right) \right); \tag{5.7}$$

3. if $\Delta = 0$,

$$y(x) = e^{-\frac{b}{2a}x}\, (c_1 x + c_2)\,. \tag{5.8}$$

5.3 Sturm separation Theorem

When a differential equation has variable coefficients, we generally cannot find explicit solutions, and this situation is pervasive in studying special functions. However, we can understand some of the qualitative properties of solutions of such equations, such as the solution's location of the zeros. We begin with a proposition about the separation of the zeros.

Theorem 5.1. Suppose that y_1 and y_2 are a fundamental pair of solutions of the differential equation (in normal form) $y' + I(x)y = 0$. If x_1 and x_2 are two consecutive zeros of y_1, then y_2 has exactly one zero in (x_1, x_2).

Proof. Assuming, by contradiction, that y_2 does not have a zero on (x_1, x_2), we also infer that y_2 does not have a zero on $[x_1, x_2]$ since y_1 and y_2 are linearly independent). Define ψ on $[x_1, x_2]$ as:

$$\psi(x) = \frac{y_1(x)}{y_2(x)}\,.$$

We observe that

1. ψ is continuous on $[x_1, x_2]$,
2. ψ' is well defined on (x_1, x_2),
3. $\psi(x_1) = \psi(x_2) = 0$.

Rolle's theorem implies there exists c with $x_1 < c < x_2$ such that $\psi'(c) = 0$. But this means the Wronskian is zero at c, which is impossible. Therefore, y_2 has at least one zero on (x_1, x_2). We claim that y_2 does not have more than one zero in (x_1, x_2). If the claim were not true, then there would exist at least two zeroes of y_2 in (x_1, x_2). Without loss of generality, let x_3 and x_4 be two consecutive zeroes of y_2 in (x_1, x_2). So we have $x_1 < x_3 < x_4 < x_2$. Then by previous arguments, y_1 will have a zero in (x_3, x_4) thus contradicting the fact that x_1 and x_2 are two consecutive zeros of y_1. □

Remark. Theorem 5.1 does not address the existence of zeros. For example, consider the equation $y'' - y = 0$, which is solved by $y(x) = \cosh x > 0$ for any $x \in \mathbb{R}$.

The behaviour of the solutions of the equation leading to hyperbolic functions $y'' - y = 0$ is not particular; when an equation is in normal form if the sign of the coefficient of y is less than or equal to zero, the solutions have at most one zero.

Theorem 5.2. If $I(x) \leq 0$ for any x in an interval $\mathcal{I}$, then no non-trivial solution of $y'' + I(x)y = 0$ can have two zeros on $\mathcal{I}$.

Proof. With no loss of generality we can assume that x_1 and x_2 are two consecutive zeros of y and that $y > 0$ in (x_1, x_2). In this situation, we have

$$y'(x_1) > 0 \quad \text{and} \quad y'(x_2) < 0. \tag{5.9}$$

In fact $y'(x_1) \leq 0$ is not possible: for, if $y'(x_1) = 0$ then, since $y(x_1) = 0$, it follows $y = 0$ identically, a contradiction. On the other hand, if $y'(x_1) < 0$ then y will be negative for $x > x_1$ and x near-by x_1 contradicting again the assumption that $y > 0$ on (x_1, x_2). The proof of the second inequality in (5.9) is similar. Consequently

$$y''(x) = -I(x)y(x) \geq 0 \quad \text{in } (x_1, x_2). \tag{5.10}$$

Inequality (5.10) implies that y' is an increasing function on (x_1, x_2). Hence

$$0 > y'(x_2) > y'(x_1) > 0.$$

This is a contradiction; thus, the non-trivial solution y can not have two zeros. □

We state the Sturm comparison Theorem.

Theorem 5.3. Let y_1 and y_2 non trivial solutions of

$$y' + I_1(x)y = 0 \quad \text{and} \quad y' + I_2(x)y = 0, \quad x \in \mathcal{I}$$

respectively. Assume I_1 and I_2 are continuous functions such that $I_1(x) \leq I_2(x)$ on $\mathcal{I}$. Then between any two consecutive zeros x_1 and x_2 of y_1, there exists at least one zero of y_2 unless $I_1(x) = I_2(x)$ on (x_1, x_2).

Proof. Let $x_1 < x_2$ be consecutive zeros of y_1. With no loss of generality, we may assume $y_1 > 0$ on (x_1, x_2). If not, consider $-y_1$. Repeating the same argument of Theorem 5.2, we can infer $y'(x_1) > 0$ and $y'(x_2) < 0$. Suppose, by contradiction, that y_2 does not have a zero on (x_1, x_2). With no loss of generality, we may assume $y_2 > 0$ on (x_1, x_2). Multiplying the equation satisfied by y_1, with y_2, and vice versa, and then subtracting the two equations we get:

$$y_1''y_2 - y_1y_2'' + (I_1(x) - I_2(x))\, y_1y_2 = 0.$$

We can rewrite the last equation as follows:

$$(y_1'y_2 - y_1y_2')' = -\left(I_1(x) - I_2(x)\right) y_1y_2.$$

Integrating on both sides from x_1 to x_2, we get

$$\left[y_1'y_2 - y_1y_2'\right]_{x_1}^{x_2} = -\int_{x_1}^{x_2} \left(I_1(x) - I_2(x)\right) y_1(x)y_2(x)\,\mathrm{d}x. \tag{5.11}$$

The left-hand side of (5.11) is non-positive, while the right-hand side is strictly positive unless $I_1(x) = I_2(x)$ on (x_1, x_2). Therefore, if $I_1(x) \neq I_2(x)$ on (x_1, x_2), we arrive at a contradiction. This finishes the proof of the theorem. □

Example. Using the separation theorem, we can infer that, for $x > 1$, the differential equation $y'' + xy = 0$ solutions are oscillatory. It is enough to observe that solutions of $y'' + y = 0$ are indeed oscillatory. Notice that with oscillatory, we mean that any solution of the equation has infinitely many zeros.

Remark. If $I(x) \geq \omega^2 > 0$, then any solution of $y'' + I(x)y = 0$ must vanish between two successive zeros of any given solution of the differential equation $y'' + \omega^2 y = 0$, which is of the form $s(x) = A\cos(\omega\, x + \varphi)$, hence in any interval of length π/ω.

5.4 Change of independent variable

Example. Consider the Cauchy-Euler differential equation:

$$a\,x^2\,y'' + b\,x\,y' + c\,y = 0\,. \tag{5.12}$$

To solve (5.12), we can change the independent variable from x to a new variable ξ. This means that we take a $\mathcal{C}^2$ bijective function $\phi : \mathbb{R} \to \mathbb{R}^+$ used for the variable transformation $x = \phi(\xi) \iff \xi = \phi^{-1}(x)$. To solve (5.12) we use the exponential function $x = e^\xi \iff \xi = \ln x$. Hence, we will obtain a transformed equation in a new dependent variable $z(\xi) = y(x) = y(e^\xi)$, or, in other terms:

$$z(\ln x) = y(x). \tag{5.13}$$

We now impose that function $y(x)$ introduced in (5.13) is indeed a solution to (5.12); with this aim in mind, we compute the first and second derivative in (5.13), using the chain rule:

$$y'(x) = \frac{1}{x} z'(\ln x), \tag{5.14}$$

$$y''(x) = -\frac{1}{x^2} z'(\ln x) + \frac{1}{x^2} z''(\ln x). \tag{5.15}$$

Inserting (5.13), (5.14), and (5.15) in (5.12) we get:

$$ax^2\left(\frac{1}{x^2}z''(\ln x)-\frac{1}{x^2}z'(\ln x)\right)+bx\frac{1}{x}z'(\ln x)+x\,z(\ln x)=0.$$

After a few calculations, we obtain

$$a\,z''(\ln x)+(b-a)\,z'(\ln x)+c\,z(\ln x)=0.$$

The last equation is, as a matter of fact, a constant coefficient equation in the new independent variable $\xi=\ln x$, namely:

$$a\,z''(\xi)+(b-a)\,z'(\xi)+c\,z(\xi)=0. \tag{5.16}$$

Once (5.16) is solved, returning to the original variable, the Cauchy Euler equation (5.12) is solved.

Consider once more the equation (5.1): with the general change of independent variables that we now represent with

$$z(\xi)=z(\varphi(x))=y(x) \tag{5.17}$$

or what is the same

$$x=\varphi^{-1}(\xi)\iff\xi=\varphi(x). \tag{5.17'}$$

Thus, the general transformation rules are

$$y(x)=z(\varphi(x)) \tag{5.18a}$$
$$y'(x)=\varphi'(x)\,z(\varphi(x)) \tag{5.18b}$$
$$y''(x)=\left[\varphi'(x)\right]^2z''(\varphi(x))+\varphi''(x)z'(\varphi(x)). \tag{5.18c}$$

Inserting (5.18) we obtain the transformation of (5.1) (which we put in explicit form)

$$z''(\varphi(x))+\frac{p(x)\varphi'(x)+\varphi''(x)}{\left[\varphi'(x)\right]^2}z'(\varphi(x))+\frac{q(x)}{\left[\varphi'(x)\right]^2}z(\varphi(x))=0 \tag{5.19}$$

Notice that the coefficients of the new linear differential operator in (5.19) are expressed as functions of the (old) variable x. They must be converted to the new ξ variable using $x=\varphi^{-1}(\xi)$ to complete the variable transformation.

Example. Consider the following differential equation:

$$(1+x^2)^2y''+2x(1+x^2)y'+4y=0.$$

We use here the change of variable $x=\tan\xi\iff\xi=\arctan x$. Being $z(\arctan x)=y(x)$ we obtain

$$y'(x)=\frac{1}{1+x^2}z'(\arctan x),$$

$$y''(x)=-\frac{2x}{(1+x^2)^2}z'(\arctan x)+\frac{1}{(1+x^2)^2}z''(\arctan x).$$

Hence, we obtain (we drop the argument of z and its derivatives)

$$(1+x^2)^2\left(-\frac{2x}{(1+x^2)^2}z' + \frac{1}{(1+x^2)^2}z''\right) + 2x(1+x^2)\frac{1}{1+x^2}z' + 4z = 0.$$

After some straightforward simplifications, we arrive at $z'' + 4z = 0$, which is easily solved by $z(\xi) = c_1\cos(2\xi) + c_2\sin(2\xi)$. Going back to the original variable, we obtain

$$y(x) = c_1\cos(2\arctan x) + c_2\sin(2\arctan x).$$

Eventually, we notice that it is possible to express $y(x)$ without trigonometric functions, in fact keeping in mind the relations:

$$\cos(2\alpha) = \frac{1-\tan^2\alpha}{1+\tan^2\alpha}, \quad \sin(2\alpha) = \frac{2\tan\alpha}{1+\tan^2\alpha},$$

we can express the general solution algebraically:

$$y(x) = \frac{1}{1+x^2}\left(c_1\,(1-x^2) + c_2\,x\right).$$

5.5 Power series representation

When the functions $p(x)$ and $q(x)$ depend on the x, very rarely solutions to (5.1) can be represented with elementary functions, i.e. polynomials, irrational functions, trigonometric, hyperbolic, exponential and logarithmic functions. The approach that will lead us to discover families of functions that are not of any of the types we have just listed is to look for solutions in the form of a power series.

Solution at ordinary points

Definition. Referring to a homogeneous linear differential equation of the form (5.1) assuming p, q analytic functions around the point x_0, we say that x_0 is an ordinary point. If either of these functions is not analytic at x_0, the x_0 is a singular point of (5.1).

Clearly, since we can use a translation $x' = x - x_0$ if necessary, we can assume $x_0 = 0$ in what follows without any loss of generality. Solutions of (5.1) in an interval containing the origin are analytic functions: the power series coefficients are individuated by a five-step procedure, known as the power series method or method of undeterminate coefficients.

Step 1. Evaluate the derivatives of the power series

$$y = \sum_{n=0}^{\infty} a_n x^n \tag{5.20a}$$

$$y' = \sum_{n=0}^{\infty} n a_n x^{n-1} \tag{5.20b}$$

$$y'' = \sum_{n=0}^{\infty} n(n-1) a_n x^{n-2}. \tag{5.20c}$$

Step 2. Collect powers of x and set the coefficient of each power equal to zero.
Step 3. The equation obtained by setting the coefficients of x^n to zero in Step 2. will contain a_j terms for a finite number of j's. Solve the equation for the a_j term having the largest index. The resulting equation is the recurrence formula for the given differential equation.
Step 4. Use the recurrence to determine a_j, $j = 2, 3, \ldots$ in terms of a_0 and a_1.
Step 5. Insert the coefficients determined in Step 4 into (5.20a) to express the solution of the differential equation (5.1).

We illustrate these five steps with some working examples. The first one is straightforward in computation but represents the procedure just described.

Example. Consider the equation also discussed in [95] Section 107:

$$y'' + 4y = 0. \tag{5.21}$$

We have chosen a straightforward problem to begin our discussion. It illustrates the procedure's steps without addressing major computational issues. The method of undeterminate coefficients is helpful in equations with variable coefficients, which we will examine later.

Step 1. Inserting the derivatives of the power series representation (5.20) in (5.21), we obtain

$$y'' + 4y = \sum_{n=0}^{\infty} n(n-1) a_n x^{n-2} + 4 \sum_{n=0}^{\infty} a_n x^n = 0. \tag{S1}$$

Step 2. Change the indexes of the terms in the second series on the right-hand side of (S1) so that the series will involve the power x^{n-2} in the general term; thus (S1) becomes

$$y'' + 4y = \sum_{n=0}^{\infty} n(n-1) a_n x^{n-2} + 4 \sum_{n=2}^{\infty} a_{n-2} x^{n-2} = 0. \tag{S2}$$

Step 3. Now add the two series in (S2) to obtain

$$y'' + 4y = \sum_{n=2}^{\infty} \Big(n(n-1)a_n + 4a_{n-2} \Big) x^{n-2} = 0. \tag{S3}$$

Summation starts from $n = 2$ because the first two terms in the first series in (S2) are zero.
Step 4. Since condition $n(n-1)a_n + 4a_{n-2} = 0$ holds, solving for a_n we obtain, for $n \geq 2$

$$a_n = -\frac{4}{n(n-1)} a_{n-2}$$

thus a_0 and a_1 are free and arbitrary. So to start the recursion fix $a_0, a_1 \in \mathbb{R}$ so that

$$\begin{aligned} a_2 &= -\frac{4}{2 \cdot 1} a_0 & a_3 &= -\frac{4}{3 \cdot 2} a_1 \\ a_4 &= -\frac{4}{4 \cdot 3} a_2 & a_5 &= -\frac{4}{5 \cdot 4} a_3 \\ a_6 &= -\frac{4}{6 \cdot 5} a_4 & a_7 &= -\frac{4}{7 \cdot 6} a_5 \end{aligned}$$

in general

$$a_{2k} = -\frac{4}{2k \cdot (2k-1)} a_{2k-2} \qquad a_{2k+1} = -\frac{4}{(2k+1) \cdot 2k} a_{2k-1}.$$

Multiplying side by side the relations of the even index, we obtain

$$a_2 \cdot a_4 \cdot \dots \cdot a_{2k} = (-1)^k \frac{4^k}{(2k)!} a_0 \cdot a_2 \cdot \dots \cdot a_{2k-2}$$

which simplifies to

$$a_{2k} = (-1)^k \frac{4^k}{(2k)!} a_0, \; k \geq 1 .$$

A similar argument for odd indexes leads to

$$a_{2k+1} = (-1)^k \frac{4^k}{(2k+1)!} a_1, \;\; k \geq 1$$

Step 5. Now recalling the assumed series (5.20a) for y we rewrite it in the form

$$\begin{aligned} y =& a_0 + \sum_{k=1}^{\infty} a_{2k} x^{2k} + a_1 x + \sum_{k=1}^{\infty} a_{2k+1} x^{2k+1} \\ =& a_0 \left(1 + \sum_{k=1}^{\infty} \frac{(-1)^k 4^k}{(2k)!} x^{2k} \right) + a_1 \left(x + \sum_{k=1}^{\infty} \frac{(-1)^k 4^k}{(2k+1)!} x^{2k+1} \right) \\ =& a_0 \sum_{k=0}^{\infty} \frac{(-1)^k}{(2k)!} (2x)^{2k} + \frac{a_1}{2} \sum_{k=0}^{\infty} \frac{(-1)^k}{(2k+1)!} (2x)^{2k+1} \\ =& a_0 \cos(2x) + a_1 \sin(2x). \end{aligned}$$

In the last step, we recognised the trigonometric functions of sine and cosine power series.

As we already pointed out, this example, despite its simplicity, is instructive in understanding how the solution-finding method works. The fact that the solution to the particular equation considered can be easily obtained with other methods allows us to check the validity of this new approach. Now, let's see how to use the power series method to solve an equation with variable coefficients.

Example. Consider, near the ordinary point $x_0 = 0$, equation

$$(1-x^2)y'' - 6xy' - 4y = 0.$$

The first difference with the previous example is that the coefficients are analytical for $|x| < 1$, so here, the power series solution (5.20a) is considered for $|x| < 1$. Inserting (5.20a), (5.20b), and (5.20c) into the equation we obtain

$$\sum_{n=0}^{\infty} n(n-1)a_n x^{n-2} - \sum_{n=0}^{\infty} n(n-1)a_n x^n - \sum_{n=0}^{\infty} 6na_n x^n - \sum_{n=0}^{\infty} 4a_n x^n = 0.$$

Now we group the identical powers, getting

$$\sum_{n=0}^{\infty} n(n-1)a_n x^{n-2} - \sum_{n=0}^{\infty} (n^2+5n+4)a_n x^n = 0,$$

or

$$\sum_{n=0}^{\infty} n(n-1)a_n x^{n-2} - \sum_{n=0}^{\infty} (n+1)(n+4)a_n x^n = 0.$$

So, rescaling the indexes in the second series, we obtain the following:

$$\sum_{n=2}^{\infty} n(n-1)a_n x^{n-2} - \sum_{n=2}^{\infty} (n-1)(n+2)a_{n-2} x^{n-2} = 0.$$

Therefore for $n \geq 2$ we have

$$n(n-1)a_n - (n-1)(n+2)a_{n-2} = 0 \implies a_n = \frac{n+2}{n} a_{n-2}.$$

Of course, a_0 and a_1 remain, as the theory requires, arbitrary since

$$n = 0 \implies 0 \cdot a_0 = 0, \quad n = 1 \implies 0 \cdot a_1 = 0.$$

We examine the recurrences obtained equating equal powers

$$\begin{aligned}
a_2 &= \frac{4}{2}\, a_0 & a_3 &= \frac{5}{3}\, a_1 \\
a_4 &= \frac{6}{4}\, a_2 & a_5 &= \frac{7}{5}\, a_3 \\
a_6 &= \frac{8}{6}\, a_4 & a_7 &= \frac{9}{7}\, a_5. \\
&\dots
\end{aligned}$$

In general

$$a_{2k} = \frac{2k+2}{2k} a_{2k-2}, \quad a_{2k+1} = \frac{2k+3}{2k+1} a_{2k-1}.$$

Thus, multiplying the "even" column we obtain

$$a_2 a_4 \dots a_{2k} = \frac{4 \cdot 6 \cdot \dots \cdot (2k+2)}{2 \cdot 4 \cdot \dots \cdot 2k} a_0 a_2 \dots a_{2k-2},$$

which simplifies in

$$a_{2k} = \frac{2k+2}{2} a_0 = (k+1)\, a_0, \;\; k \geq 1.$$

For even terms, a similar argument leads to:

$$a_{2k+1} = \frac{2k+3}{3} a_1, \;\; k \geq 1.$$

Now, starting from the standard representation:

$$y = \left(a_0 + \sum_{k=1}^{\infty} a_{2k} x^{2k} \right) + \left(a_1 x + \sum_{k=1}^{\infty} a_{2k+1} x^{2k+1} \right),$$

it follows that

$$y = a_0 \left(1 + \sum_{k=1}^{\infty} (k+1) x^{2k} \right) + a_1 \left(x + \sum_{k=1}^{\infty} \frac{2k+3}{3} x^{2k+1} \right).$$

It is possible to verify with the usual convergence tests that both series converge for $|x| < 1$.

Finally, we observe that both power series are summed in terms of elementary functions since they are related to the geometric series:

$$y = a_0 \sum_{k=0}^{\infty} (k+1) x^{2k} + a_1 \sum_{k=0}^{\infty} \frac{2k+3}{3} x^{2k+1} = \frac{a_0}{(1-x^2)^2} + \frac{a_1}{3} \frac{3x - x^3}{(1-x^2)^2}.$$

Remark. In chapter 6, we will present two classes of special functions, the Airy functions and the Hermite polynomials, obtained from equations with ordinary points.

Solution at regular singular points

The search method for indeterminate coefficients just presented cannot be used around the points where the coefficients $p(x)$ and $q(x)$ of the differential operator L have singularities. However, as we will see, it is still possible to adapt this method to allow for the search for solutions in the presence of some specific singularities.

Definition. When, given the linear differential operator (5.1), there exists x_0 such that $p(x)$ and $q(x)$ are not analytic in x_0, we say that x_0 is a singular point. The singular point x_0 is a *regular* singular point of equation (5.1) if both functions $(x - x_0)\, p(x)$ and $(x - x_0)^2 q(x)$ are analytic at x_0. Singular points that are not regular are called *irregular.*

Once again, barring translations, we can limit ourselves to illustrating the procedure for finding solutions by assuming that $x_0 = 0$ is a regular singular point of (5.1). The power series method has to be modified following the so-called Frobenius method. Here, we limit ourselves to a brief mention; see, for instance, [34] or [95] for extensive treatment. In the presence of regular singular points, we search for power series solutions of the form

$$y = \sum_{n=0}^{\infty} a_n x^{n+\alpha} \tag{5.22}$$

where the constant α will be determined by solving a quadratic equation, called *indicial equation.* We illustrate the procedure with an example.

Example. Consider equation

$$2xy'' + (1 + x)y' - 2y = 0. \tag{5.23}$$

Here, $x_0 = 0$ is a regular singular point for (5.23). We search for a solution to the form (5.22). Differentiating we obtain

$$y' = \sum_{n=0}^{\infty} (n + \alpha) a_n x^{n+\alpha-1} \tag{5.24a}$$

$$y'' = \sum_{n=0}^{\infty} (n + \alpha)(n + \alpha - 1) a_n x^{n+\alpha-2}\,. \tag{5.24b}$$

Thus inserting (5.22), (5.24a), and (5.24b) in (5.23), we get some simplifications:

$$\sum_{n=0}^{\infty} (n + \alpha)(2n + 2\alpha - 1) a_n x^{n+\alpha-1} + \sum_{n=0}^{\infty} (n + \alpha - 2) a_n x^{n+\alpha}\,. \tag{5.25}$$

To individuate α we equate the coefficient of the lowest power of x in (5.25) to zero, here $x^{\alpha-1}$, obtaining in this way the *indicial equation*:

$$2\alpha(\alpha - 1)\, a_0 + \alpha\, a_0 = 0. \tag{5.26}$$

Solving (5.26), after eliminating a_0 which we assume non-zero, we obtain the indicial roots $\alpha = 0$ and $\alpha = \frac{1}{2}$. Equating to zero the powers of $x^{n+\alpha}$, we obtain the recurrence rule for the coefficient of the power series solution to (5.23):

$$(n + \alpha)(2n + 2\alpha - 1) a_n + (n + \alpha - 3) a_{n-1} = 0\,. \tag{5.27}$$

We start with $\alpha = 0$. Form (5.27) we obtain the recursion relation

$$a_n = \frac{3-n}{n(2n-1)}\, a_{n-1}$$

which gives

$$\begin{aligned} n &= 1 \qquad a_1 = 2a_0 \\ n &= 2 \qquad a_2 = \frac{1}{6}\, a_1 = \frac{1}{3}\, a_0 \\ n &\geq 2 \qquad a_n = 0\,. \end{aligned}$$

So we found a polynomial solution $y_1 = a_0 \left(1 + 2x + \frac{1}{3}\, x^2\right)$. If $\alpha = \frac{1}{2}$ (5.27) gives

$$a_n = -\frac{(2n-5)}{2n(2n+1)}\, a_{n-1}$$

which, for the first values of n, gives

$$\begin{aligned} n = 1 \to \quad a_1 &= \frac{3}{2\cdot 1\cdot 3} a_0 \\ n = 2 \to \quad a_2 &= \frac{1}{2\cdot 2\cdot 5}\, a_1 \\ n = 3 \to \quad a_3 &= \frac{-1}{2\cdot 3\cdot 7}\, a_2 \\ n = 4 \to \quad a_4 &= \frac{-3}{2\cdot 4\cdot 9}\, a_3 \\ n = 5 \to \quad a_5 &= \frac{-5}{2\cdot 5\cdot 11}\, a_4 \\ n = 6 \to \quad a_6 &= \frac{-7}{2\cdot 6\cdot 13}\, a_5\,. \end{aligned}$$

This suggests that the general term is expressed by

$$a_n = -\frac{(2n-5)}{2n(2n+1)} a_{n-1}.$$

The column product then yields, for $n \geq 1$

$$a_n = \frac{(-1)^n 3a_0}{2^n n!(2n-3)(2n-1)(2n+1)}.$$

Thus, we can write the second solution of (5.23) as:

$$y_2 = x^{1/2} + \sum_{n=1}^{\infty} \frac{(-1)^n 3a_0}{2^n n!(2n-3)(2n-1)(2n+1)}\, x^{n+1/2}.$$

This example demonstrates that elementary functions alone are not enough to represent the solutions of this class of differential equations: in this case, the solution just determined has a power series development that is not attributable to that of trigonometric, exponential, and logarithmic functions.

Chapter 6

Airy and Hermite functions

Starting from this chapter, after the premises on linear differential equations of order two, we can study the special functions that originate from these equations in cases where they do not admit elementary solutions. We begin the discussion by presenting two classes of functions, Hermite and Airy functions, which are generated by equations with variable coefficients with ordinary points. The following chapters are dedicated to equations with singularities.

6.1 Airy differential equation

When the solutions of a differential equation cannot be expressed in terms of elementary functions, the search for solutions in the form of a power series becomes the primary method of attack on the problem if the coefficients of the differential operator are analytic functions. Here, we expose how the power series method works for the Airy equation, which is, in terms of its formulation, the most straightforward second-order linear differential equation with variable coefficients:

$$.y'' - xy = 0 \tag{6.1}$$

However, the power-series solution process is much more demanding than the examples in the previous chapter. Let's start the usual power series procedure by inserting (5.20a) and (5.20c) into the Airy equation (6.1)

$$\sum_{n=0}^{\infty} n(n-1)a_n x^{n-2} - \sum_{n=0}^{\infty} a_n x^{n+1} = 0\,. \tag{6.1a}$$

From (6.1a) we immediately infer that $a_2 = 0$ and then we rewrite (6.1a) as

$$\sum_{n=3}^{\infty} n(n-1)a_n x^{n-2} - \sum_{n=0}^{\infty} a_n x^{n+1} = 0. \tag{6.1b}$$

Now we rearrange the indexes in the first series in (6.1b) to obtain

$$\sum_{n=0}^{\infty} \Big((n+3)(n+2)a_{n+3} - a_n\Big)x^{n+1} = 0. \tag{6.1c}$$

DOI: 10.1201/9781003171546-6

Thus the sequence (a_n) of the coefficients of the power series solution of equation (6.1) verifies condition

$$\begin{cases} a_{n+3} = \dfrac{1}{(n+3)(n+2)}\, a_n \\ a_0,\, a_1,\, a_2 \in \mathbb{R}, \text{ with } a_2 = 0. \end{cases} \tag{6.2}$$

Equation (6.2) identifies a recurrence of the third order, which means that three indexed families a_{3n}, a_{3n+1}, a_{3n+2} are generated by (6.2) starting from a_0, a_1, a_2. Since we saw that $a_2 = 0$, we see that $a_{3n+2} = 0$ for any $n \in \mathbb{N}$, while a_{3n} and a_{3n+1} are determined as follows:

$$\begin{aligned} a_3 &= \frac{1}{6 \cdot 5}\, a_0 & a_4 &= \frac{1}{7 \cdot 6}\, a_1 \\ a_6 &= \frac{1}{9 \cdot 8}\, a_3 & a_7 &= \frac{1}{10 \cdot 9}\, a_4 \\ a_9 &= \frac{1}{12 \cdot 11}\, a_6 & a_{10} &= \frac{1}{13 \cdot 12}\, a_7. \\ &\dots \end{aligned}$$

In general

$$a_{3n} = \frac{1}{3n(3n-1)} a_{3n-3}, \qquad a_{3n+1} = \frac{1}{(3n+1)3n} a_{3n-2}.$$

Thus, multiplying by columns, we obtain, for $n \geq 1$:

$$a_{3n} = \frac{a_0}{\displaystyle\prod_{k=1}^{n} 3k(3k-1)}, \qquad a_{3n+1} = \frac{a_1}{\displaystyle\prod_{k=1}^{n} (3k+1)3k}. \tag{6.3}$$

Both products in (6.3) are computed in closed form using the Euler Gamma function; in fact, first write product which define a_{3n} as:

$$\prod_{k=1}^{n} 3k(3k-1) = 3^{2n} n! \prod_{k=1}^{n} \left(k - \frac{1}{3}\right).$$

Then observe that

$$\begin{aligned} \prod_{k=1}^{n} \left(k - \frac{1}{3}\right) &= \left(n - \frac{1}{3}\right)\left(n - 1 - \frac{1}{3}\right) \dots \left(1 - \frac{1}{3}\right) \\ &= \left(n - 1 + \frac{2}{3}\right)\left(n - 2 + \frac{2}{3}\right) \dots \left(\frac{2}{3}\right) \end{aligned}$$

we can invoke equations (2.12) and (2.13), see page 62 to arrive at

$$\prod_{k=1}^{n} \left(k - \frac{1}{3}\right) = \frac{\Gamma\left(n + \frac{2}{3}\right)}{\Gamma\left(\frac{2}{3}\right)}.$$

So that

$$\prod_{k=1}^{n} 3k(3k-1) = \frac{3^{2n} n!\,\Gamma\left(n+\frac{2}{3}\right)}{\Gamma\left(\frac{2}{3}\right)}. \tag{6.4}$$

A similar argument leads to

$$\prod_{k=1}^{n} 3k(3k+1) = \frac{3^{2n} n!\,\Gamma\left(n+\frac{4}{3}\right)}{\Gamma\left(\frac{4}{3}\right)}. \tag{6.5}$$

Now, starting from the representation

$$y = \left(a_0 + \sum_{k=1}^{\infty} a_{3n} x^{3nk}\right) + \left(a_1 x + \sum_{k=1}^{\infty} a_{3n+1} x^{3n+1}\right),$$

from (6.4) and (6.5) follows that

$$y = a_0 \left(1 + \sum_{n=1}^{\infty} \frac{\Gamma\left(\frac{2}{3}\right)}{3^{2n} n!\,\Gamma\left(n+\frac{2}{3}\right)} x^{3n}\right) + a_1 \left(x + \sum_{n=1}^{\infty} \frac{\Gamma\left(\frac{4}{3}\right)}{3^{2n} n!\,\Gamma\left(n+\frac{4}{3}\right)} x^{3n+1}\right),$$

so that the general solution of (6.1) is

$$y = a_0 \Gamma\left(\frac{2}{3}\right) \sum_{n=0}^{\infty} \frac{1}{3^{2n} n!\,\Gamma\left(n+\frac{2}{3}\right)} x^{3n} + a_1 \Gamma\left(\frac{4}{3}\right) \sum_{n=0}^{\infty} \frac{1}{3^{2n} n!\,\Gamma\left(n+\frac{4}{3}\right)} x^{3n+1}. \tag{6.6}$$

Remark. In the study of Optics questions, the equation (6.1) was first studied by Airy. Using integration techniques on a suitable contour in the complex plane, he obtained the following integral representations (6.7) below, which define a fundamental system of solutions:

$$\mathrm{Ai}(x) = \frac{1}{\pi} \int_0^{+\infty} \cos\left(x\,u + \frac{u^3}{3}\right) du \tag{6.7a}$$

$$\mathrm{Bi}(x) = \frac{1}{\pi} \int_0^{+\infty} \left[\cos\left(x\,u + \frac{u^3}{3}\right) + \exp\left(x\,u - \frac{u^3}{3}\right)\right] du. \tag{6.7b}$$

We refer the interested reader to the proof of equations (6.7a) and (6.7b) in the monograph [110], which is entirely devoted to such functions and provides several applications of the Airy functions.

We end our brief presentation on Airy functions by highlighting two interesting facts. First of all, the two power series identified in (6.6) allow, by the results of existence and uniqueness, to represent the Airy functions (6.7a) and (6.7b): for instance, referring to (6.6), the function Ai in (6.7a) is represented by choosing:

$$a_0 = \frac{1}{3^{2/3}\Gamma\left(\frac{2}{3}\right)}, \tag{6.8a}$$

$$a_1 = -\frac{1}{3^{1/3}\Gamma\left(\frac{1}{3}\right)}. \tag{6.8b}$$

This last statement stems from direct computations, using first the variable change $\frac{u^3}{3} = v$ and then recalling formula (2.8aa), see page 61:

$$\mathrm{Ai}(0) = \frac{1}{\pi} \int_0^{+\infty} \cos\left(\frac{u^3}{3}\right) du = \frac{1}{\pi\, 3^{\frac{2}{3}}} \cos\tfrac{\pi}{6}\, \Gamma\left(\tfrac{1}{3}\right) .$$

Thus, invoking the reflection formula (2.32), we arrive at

$$\mathrm{Ai}(0) = \frac{1}{3^{\frac{2}{3}} \Gamma\left(\frac{2}{3}\right)} = a_0 .$$

In the same way, but this time using the formula (2.8b'), we obtain

$$\mathrm{Ai}'(0) = -\frac{1}{\pi} \int_0^{+\infty} u \sin\left(\frac{u^3}{3}\right) du = -\frac{1}{\pi\, 3^{1/3}} \sin\tfrac{\pi}{3}\, \Gamma\left(\tfrac{2}{3}\right) .$$

Eventually recalling again (2.32), we arrive at

$$\mathrm{Ai}'(0) = -\frac{1}{3^{\frac{1}{3}}\, \Gamma(\frac{1}{3})} .$$

The second consideration regarding Airy functions concerns what we will obtain in the next section when we express the general solution of equation (6.1) in terms of Bessel functions, which are a particular type of hypergeometric functions.

6.2 Hermite polynomials

The Hermite differential equation is one one-parameter family of linear second-order differential equations of the form

$$y'' - 2x\, y' + 2\nu\, y = 0 \tag{6.9}$$

being $\nu \in \mathbb{C}$. In the following, we refer to (6.9) as the Hermite equation of order ν. It is clear that $x = 0$ is an ordinary point of (6.9); hence we can replicate with no difficulties the procedure illustrated in Section 5.5, to produce the power series solution where, in this case, the coefficient sequence $(a_n)_{n\in\mathbb{N}}$ fulfils $(n+2)(n+1)a_{n+2} + 2(\nu - n)a_n = 0$ or

$$a_{n+2} = -\frac{2(\nu - n)}{(n+2)(n+1)}\, a_n . \tag{6.10}$$

By dividing the solutions of the recurring equation between even and odd indices, two independent solutions are determined, identifying the general solution of the Hermite equation. Moreover, from the recurrence (6.10), it is understood a priori that the value of the parameter ν heavily influences the behaviour of the solution: in fact, if $\nu = m \in \mathbb{N}$, there exists a solution of the Hermite equation that reduces to a polynomial. In this case, we speak of Hermite polynomials. The general solution to (6.9) is, therefore given by $y = a_0 y_0 + a_1 y_1$ where

$$y_0 = 1 - \frac{2\nu}{2!}x^2 + \frac{2^2\nu(\nu-2)}{4!}x^4 + \ldots \tag{6.11a}$$

$$y_1 = x - \frac{2(\nu-1)}{3!}x^3 + \frac{2^2(\nu-1)(\nu-3)}{5!}x^5 + \ldots \tag{6.11b}$$

In chapters 7 and 10, we will place these functions and the Airy functions introduced previously in the family of Bessel and generalised hypergeometric functions of one variable. Here, we face the interesting situation of $\nu = n \in \mathbb{N}$. In this case, the differential equation (6.9) has one polynomial solution, which will be denoted with $H_n(x)$. It can be written in closed form using a single expression that does need to distinguish between the case of even degree and that of odd degree using the integer part; see also [95, page 399] or [74, page 60]:

$$H_n(x) = \sum_{k=0}^{\lfloor \frac{n}{2} \rfloor} \frac{(-1)^k n!(2x)^{n-2k}}{k!(n-2k)!}. \tag{6.12}$$

We list below the first few Hermite polynomials

$$\begin{aligned}
H_0(x) &= 1\\
H_1(x) &= 2x\\
H_2(x) &= 4x^2 - 2\\
H_3(x) &= 8x^3 - 12x\\
H_4(x) &= 16x^4 - 48x^2 + 12
\end{aligned}$$

An alternative and useful way to introduce Hermite polynomials is based on the following formula (6.13), which some authors, such as [74], use as their primary definition:

$$H_n(x) = (-1)^n e^{x^2} \frac{\mathrm{d}^n}{\mathrm{d}x^n} e^{-x^2}. \tag{6.13}$$

Through an elementary but challenging direct calculation, it is possible to verify that the two formulas (6.12) and (6.13) generate the same family of polynomials, solutions of the differential equation (6.9) with integer index n. Moreover, following again [74] we observe that the family of polynomials $H_n(x)$ can be obtained by observing that for any $t \in \mathbb{R}$

$$e^{2xt-t^2} = \sum_{n=0}^{\infty} \frac{H_n(x)}{n!} t^n. \tag{6.14}$$

For this reason, the function $g(x,t) = e^{2xt-t^2}$ is called the generating function of the Hermite polynomials. The proof of (6.14) is obtained by developing the function in power series with respect to t function $g(x,t) = e^{2xt-t^2}$. This fact is important because it allows us to discover a recurrence between the Hermite polynomials. Observing that differentiating with respect to t, we have

$$g_t(x,t) = 2(x-t)g(x,t) \tag{6.15}$$

since differentiation term by term is indeed possible for the power series on the right-hand side of (6.14), we can plug it into (6.15):

$$\sum_{n=0}^{\infty} \frac{H_n(x)}{(n-1)!} t^{n-1} = 2x \sum_{n=0}^{\infty} \frac{H_n(x)}{n!} t^n - 2 \sum_{n=0}^{\infty} \frac{H_n(x)}{n!} t^{n+1}$$

so, by rescaling some indices, we arrive at

$$H_{n+1}(x) = 2xH_n(x) + -2nH_{n-1}(x) \tag{6.16}$$

which holds for $n \in \mathbb{N}$.

A fact worth mentioning is that if in the equation with variable coefficients

$$y'' + (2n+1-x^2)y = 0 \tag{6.17}$$

we change the independent variable as $y = e^{-\frac{x^2}{2}}\, u$ we see that (6.17) is solved taking $u = H_n(x)$. For instance taking $n = 2$ we have that $y = (4x^2-2)e^{-\frac{x^2}{2}}$ is a particular solution to $y'' + (5-x^2)y = 0$. Apart from the fact, however, very interesting, given that through Hermite polynomials, it is possible to solve in terms of elementary functions the equation (6.17) from this equation, we deduce the important weighted orthogonality relation satisfied by Hermite's polynomials.

Theorem. For positive integers m, n, $n \neq m$ we have

$$\int_{-\infty}^{+\infty} e^{-x^2}\, H_m(x)\, H_n(x)\, \mathrm{d}x = 0 \tag{6.18}$$

Proof. To show (6.18) let us introduce functions $u_m = e^{-\frac{x^2}{2}} H_m$ and $u_n = e^{-\frac{x^2}{2}} H_m$ which both solve (6.17) for the appropriate index, the differentiating $u_n' u_m - u_m' u_n$ we obtain

$$\frac{\mathrm{d}}{\mathrm{d}x}\left(u_n' u_m - u_m' u_n\right) = u_m u_n'' - u_n u_m''.$$

Hence using twice (6.17), we arrive at

$$\frac{\mathrm{d}}{\mathrm{d}x}\left(u_n' u_m - u_m' u_n\right) = u_n\left(-2n + x^2 - 1\right) u_m - u_m\left(-2m + x^2 - 1\right) u_n,$$

which, after some simplifications, reduces to

$$\frac{\mathrm{d}}{\mathrm{d}x}\left(u_n' u_m - u_m' u_n\right) = 2(m-n) u_m u_n \,. \tag{6.19}$$

To end our argument, integrate both sides of (6.19) and observe that the left-hand side vanishes due to the exponential factor, proving (6.18). □

Differentiating $g(x,t)$ with respect to t, we obtain a second recursive relation since

$$g_t(x,t) = 2t\, g(x,t)$$

we can again differentiate term by term (6.14), arriving at

$$\sum_{n=1}^{\infty} \left(H_{n-1}'(x) - 2nH_{n-1}(x)\right) \frac{t^n}{n!}$$

and then, for $n \in \mathbb{N}$

$$H_n'(x) = 2nH_{n-1}(x)\,. \tag{6.20}$$

Hence comparing (6.16) and (6.20) we arrive at

$$H_{n+1}(x) - 2xH_n(x) + H_n'(x) = 0 \tag{6.21}$$

so that, differentiating (6.21) and using again (6.20), we see that H_n is indeed solution to (6.9) when $\nu = n \in \mathbb{N}$.

To obtain an integral representation of the Hermite polynomial, first, we need to compute the following parametric integral

$$\jmath(x) = \int_0^{+\infty} e^{-u^2} \cos(2u\, x)\, \mathrm{d}u. \tag{6.22}$$

Observing that the exponential decay of the integrand legitimates differentiation of (6.22), we arrive, after integration by parts, at

$$\jmath'(x) = \int_0^{\infty} -2ue^{-u^2} \sin(2u\, x)\, \mathrm{d}u = -2x\, \jmath(x).$$

Recalling that $\jmath(0) = \sqrt{\pi}/2$ we obtain

$$\jmath(x) = \frac{\sqrt{\pi}}{2}\, e^{-x^2}$$

Hence, we can use this fact to obtain a nice representation of e^{-x^2}:

$$e^{-x^2} = \frac{2}{\sqrt{\pi}} \int_0^{+\infty} e^{-u^2} \cos(2u\, x)\, \mathrm{d}u. \tag{6.23}$$

Recalling (6.13) and differentiating n-times on both sides of (6.23), we arrive at the integral representation of the Hermite polynomials. For even indexes

$$H_{2n}(x) = (-1)^n \frac{2^{2n+1}}{\sqrt{\pi}}\, e^{-x^2} \int_0^{+\infty} u^{2n} e^{-u^2} \cos(2u\,x)\,du.$$

For odd indexes

$$H_{2n+1}(x) = (-1)^n \frac{2^{2n+2}}{\sqrt{\pi}}\, e^{-x^2} \int_0^{+\infty} u^{2n+1} e^{-u^2} \sin(2u\,x)\,du.$$

Chapter 7

Bessel functions: Part 1

In 1764, Euler, investigating the vibrations of a stretched membrane, was the first mathematician to get in touch with the differential equation, now known as the Bessel equation. Later, Daniel Bernoulli, Lagrange, Laplace, and Poisson also worked on this equation. Anyway, the German mathematical astronomer Friedrich Wilhelm Bessel, to whom, among other things, the determination of the orbit of Halley's Comet is due, studied equation (7.1) below while working on dynamical astronomy. A detailed historical reconstruction is provided in [43]. Since their introduction, these functions have proven to be enormously useful, considering the great number of their applications in every scientific field. In consideration of the vastness of the subject to be treated, we have chosen to break our presentation into two chapters, a choice which, then, is made natural by the fact that the two independent solutions providing the general solution of (7.1) have, as we will see, different behaviours regarding their boundedness. Furthermore, since (7.1) is a parametric equation, this first chapter emphasises mainly integer index equations, which are relevant for many important applications.

7.1 Bessel differential equation

The Bessel equation of the first kind and order ν is the following second-order linear differential equation, where parameter ν is any number, real or complex:

$$x^2 y'' + x y' + (x^2 - \nu^2) y = 0. \tag{7.1}$$

Many useful applications relate to equations whose order is an integer or a semi-integer. The Bessel functions discussed in this chapter are particular solutions of the equation (7.1) that satisfy appropriate conditions we will specify later. To represent them, let's retrace the history of their discovery in what seems to us the natural way to tackle the definition of special functions, in this case, regarding them as solutions of a parametric family of differential equations rather than their introduction through a generating function. However, the representation of Bessel functions using a generating function, which we will detail in Section 7.7, allows us to investigate properties of considerable importance, first of all, the possibility of obtaining, in a natural way, through

DOI: 10.1201/9781003171546-7

the Fourier series, the integral representation of the Bessel functions of integer order. We begin our discussion by examining the two special functions that originated from the equations most commonly used in applications: order zero and one. We will deal with the general situation later, but given the great frequency of the use of these functions, it makes sense to give them specific emphasis. It is essential to underline that the functions we are going to introduce using the notation J_0 and J_1 are not only characterised by having an integer index but, as the whole family of functions that we will denote with J_n, they are individuated by considering the bounded solutions of (7.1) for $x \to 0^+$.

7.2 Bessel function of order zero

To become familiar with the Bessel equation, we begin our study by considering (7.1) in the particular case $\nu = 0$:

$$xy'' + y' + xy = 0. \tag{7.2}$$

This case is peculiar since the factor x can be simplified from the equation. Moreover, if we assume that the solution is $\mathcal{C}^2$ in $x_0 = 0$, which indeed is a singular regular point, we have that it must necessarily be $y'(0) = 0$.

Theorem. The solution of (7.2) such that $y(0) = 1$ is expressed by the power series:

$$J_0(x) = \sum_{n=0}^{\infty} \frac{(-1)^n}{(n!)^2} \left(-\frac{x}{2}\right)^{2n}. \tag{7.3}$$

This function is called the Bessel function of the first kind of order zero.

Remark. The term "first kind" underlines that one wants to determine among the solutions of (7.2) those bounded at the origin, which is a regular singular point of the equation. As we will soon find out when using Abel's formula (5.3), we will see that the solution of (7.2) individuated in the process, which completes the general integral of (7.2), is not bounded in the origin; for this reason, it will be called the Bessel function of the "second kind".

Proof. Following [38] and [33], we use the Laplace transform to integrate this equation. We, therefore, recall Remark 1.31, see page 54.

We already remarked that $y'(0) = 0$ follows from the nature of the equation (7.2). Moreover, if we also assume $y(0) = 1$, according to the aforementioned requirement of searching solutions of (7.2) bounded in the origin, we find:

$$\mathscr{L}\left[xy'' + y' + xy\right](s) = -\frac{\mathrm{d}}{\mathrm{d}s}\left(s^2 Y(s) - sy(0)\right) + \left(sY - y(0)\right) - \frac{\mathrm{d}}{\mathrm{d}s} Y(s).$$

After some calculations, we obtain, dropping the argument s, the separable differential equation:

$$(s^2+1)\frac{\mathrm{d}}{\mathrm{d}s}Y = -sY.$$

Thus, if c stands for the integration constant, we arrive at

$$Y(s) = \frac{c}{\sqrt{1+s^2}}.$$

To determine the inverse transform of $Y(s)$, expanding in power series in terms of $1/s$ is helpful, so we express Y changing s to $1/q$. Hence, we can use the binomial series expansions to obtain the following relation:

$$Y(1/q) = \frac{cq}{\sqrt{1+q^2}} = c\sum_{n=0}^{\infty}\binom{-1/2}{n} q^{2n+1}.$$

Going back to s, we see that

$$Y(s) = c\sum_{n=0}^{\infty}\binom{-1/2}{n} s^{-(2n+1)}.$$

Now using the inversion relation $\mathscr{L}^{-1}\left[\frac{1}{s^{n+1}}\right](x) = \frac{x^n}{n!}$ we arrive at

$$y(x) = c\sum_{n=0}^{\infty}\binom{-1/2}{n}\frac{x^{2n}}{(2n)!}$$

Finally, we observe that from $y(0) = 1$ follows $c = 1$, i.e.

$$J_0(x) = \sum_{n=0}^{\infty}\binom{-1/2}{n}\frac{x^{2n}}{(2n)!}.$$

Make the binomial coefficient explicit:

$$\binom{-1/2}{n} = (-1)^n\frac{(2n-1)!!}{2^n n!}$$

we get

$$J_0(x) = \sum_{n=0}^{\infty}(-1)^n\frac{(2n-1)!!}{2^n n!}\frac{x^{2n}}{(2n)!}.$$

Now recalling that $(2n-1)!! = \frac{(2n-1)!}{2^{n-1}(n-1)!}$ we arrive at

$$J_0(x) = \sum_{n=0}^{\infty}(-1)^n\frac{(2n-1)!}{2^{n-1}(n-1)!}\frac{1}{2^n n!}\frac{x^{2n}}{(2n)!}.$$

After simplifying, we obtain (7.3), ending our argument. □

The Laplace transform technique we just followed to solve (7.2) also allows us to deduce some interesting properties of J_0. We begin introducing the hat notation for its Laplace transform:

$$\mathscr{L}\left[J_0(x)\right](s) = \widehat{J_0}(s) = \frac{1}{\sqrt{1+s^2}}.$$

If we write the latter explicitly

$$\int_0^\infty e^{-sx} J_0(x)\,\mathrm{d}x = \frac{1}{\sqrt{1+s^2}},$$

taking $s = 0$, we see that

$$\int_0^\infty J_0(x)\,\mathrm{d}x = 1.$$

Moreover, from

$$\widehat{J_0}'(s) = \left(\frac{1}{\sqrt{1+s^2}}\right)' = -\frac{s}{(1+s^2)^{3/2}}$$

we infer

$$\int_0^\infty e^{-sx} x\,J_0(x)\,\mathrm{d}x = \frac{s}{(1+s^2)^{3/2}}.$$

Hence, taking $s = 0$ again, we see that

$$\int_0^\infty x\,J_0(x)\,\mathrm{d}x = 0,$$

which is straightforward since J_0 is an even function. So far, continuing, if we calculate the second derivative and then take $s = 0$, we obtain

$$\int_0^\infty x^2\,J_0(x)\,\mathrm{d}x = 1.$$

We now provide an integral representation of J_0.

Theorem. J_0 admits the following integral representation

$$J_0(x) = \frac{1}{\pi}\int_0^\pi \cos(x\sin\vartheta)\,\mathrm{d}\vartheta. \tag{7.4}$$

Proof. The function at right hand side of (7.4) is such that $J_0(0) = 1$. We need to show that it solves (7.2). It is not against the generality of the following argument to work with the function $j_0(x) = \pi\,J_0(x)$. Hence

$$j_0(x) = \int_0^\pi \cos(x\sin\alpha)\,\mathrm{d}\alpha, \quad j_0'(x) = -\int_0^\pi \sin\alpha\,\sin(x\sin\alpha)\,\mathrm{d}\alpha.$$

We continue by integrating by parts:

$$j_0'(x) = \Big[\cos\alpha \sin(x\sin\alpha)\Big]_{\alpha=0}^{\alpha=\pi} - x\int_0^\pi \cos^2\alpha\, \cos(x\sin\alpha)\, d\alpha$$
$$= -x\int_0^\pi \cos^2\alpha\, \cos(x\sin\alpha)\, d\alpha.$$

Moreover, by differentiating j_0 directly for its definition, we obtain

$$j_0''(x) = -\int_0^\pi \sin^2\alpha\, \cos(x\sin\alpha)\, d\alpha.$$

Therefore, (7.4) follows after some computations:

$$x j_0''(x) + j_0'(x) + x j_0(x) = -x\int_0^\pi \sin^2\alpha\, \cos(x\sin\alpha)\, d\alpha$$
$$- x\int_0^\pi \cos^2\alpha\, \cos(x\sin\alpha)\, d\alpha + x\int_0^\pi \cos(x\sin\alpha)\, d\alpha = 0.$$

□

Remark. How we obtain the integral representation (7.4) is unnatural in some way. We will see how this representation results from a suitable Fourier series expansion; see Theorem 7.10. Furthermore, it is important to note that the integral representation (7.4) tells us that J_0 is bounded on the real axes.

The following theorem is a clear example of the use of techniques typical when studying special functions. We will present a definite integration formula relating to the product of two functions obtained by dilating the argument of J_0. This integral formula is due to Lommel, who studied special functions obtained from a particular nonhomogenous Bessel equation. We need a preliminary result concerning a modified Bessel equation of order zero.

Lemma 7.1. If $\alpha \neq 0$ is a real number, then a solution to the following differential equation:

$$x\,y'' + y' + \alpha^2\, x\, y = 0 \tag{7.5}$$

is given by $J_0(\alpha\, x)$.

Proof. We use the fact that $J_0(x)$ solves (7.2) and the change of independent variable $\xi = \alpha\, x$ in (7.5). We introduce, following (5.17) and (5.18), $u(\xi) = u(\alpha\, x) = y(x)$. Clearly $y'(x) = \alpha\, u'(\xi)$ and $y''(x) = \alpha^2\, u''(\xi)$; hence, inserting in (7.5), we obtain

$$x\alpha^2\, u''(\xi) + \alpha\, u'(\xi) + \alpha^2\, x\, u(\xi) = 0.$$

Expressing x as a function of ξ we get

$$\frac{\xi}{\alpha}\alpha^2\, u''(\xi) + \alpha\, u'(\xi) + \alpha^2\, \frac{\xi}{\alpha}\, u(\xi) = 0.$$

So that, by dividing by α we arrive at

$$\xi\, u''(\xi) + u'(\xi) + \xi\, u(\xi) = 0.$$

This last equation is a Bessel equation of order zero, solved by $u(\xi) = J_0(\xi)$; hence, going back to the original variable, we see that $y(x) = J_0(\alpha\, x)$ solves (7.1) ending our argument. □

Remark. The method used in Lemma 7.1 works using any solution of (7.2), not necessarily J_0.

Theorem 7.2. If $\alpha \neq \beta$ are two real numbers, then for any $x \in \mathbb{R}$, we have

$$\int_0^x s\, J_0(\alpha\, s)\, J_0(\beta\, s)\, \mathrm{d}s = \frac{x}{\beta^2 - \alpha^2}\Big(\alpha\, J_0'(\alpha\, x) J_0(\beta\, x) - \beta\, J_0'(\beta\, x) J_0(\alpha\, x)\Big). \tag{7.6}$$

Remark. In this case, the thesis (7.6) does not hold for arbitrary solutions of the equation (7.2) but exclusively for J_0.

Proof. In force of Lemma 7.1 we have that $u(x) = J_0(\alpha\, x)$ and $v(x) = J_0(\beta\, x)$ are respectively solutions of

$$x\, u'' + u' + \alpha^2\, u = 0, \tag{7.7a}$$

$$x\, v'' + v' + \beta^2\, v = 0. \tag{7.7b}$$

Following [24] chapter 1, Section 10 we multiply (7.7a) for v and (7.7b) for u and then we subtract side by side arriving at

$$x\,(u''v - uv'') + (u'v - uv') = \left(\beta^2 - \alpha^2\right) x\, uv\,.$$

We rewrite the left-hand side of the last equality, obtaining:

$$\frac{\mathrm{d}}{\mathrm{d}x}\big[x\,(u'v - uv')\big] = \left(\beta^2 - \alpha^2\right) x\, uv\,. \tag{7.8}$$

Thus, choosing a real x and integrating (7.8) in $[0, x]$, recalling that $u(x) = J_0(\alpha\, x)$, $v(x) = J_0(\beta\, x)$ and that $J_0(0) = 0$ we arrive at (7.6) ending the proof of Theorem 7.2. □

An immediate consequence of (7.6) is expressed by the following result:

Corollary 7.3. If $\alpha \neq \beta$ are two distinct roots of $J_0(x) = 0$, then

$$\int_0^1 s\, J_0(\alpha\, s)\, J_0(\beta\, s)\, \mathrm{d}s = 0.$$

Remark. In Section 7.4, we will show that equation $J_0(x) = 0$ has infinitely many roots. Moreover thesis of Corollary 7.3 means orthogonality between functions $J_0(\alpha\, s)$ and $J_0(\beta\, s)$. As a matter of fact, it is possible to develop a Fourier series theory known as Fourier Bessel expansion. At our introductory level, we will not treat this argument here, which is discussed in great detail in the monumental treatise [113].

From the point of view of the theory of differential equations, the determination of J_0 allows completing the general solution of (7.2) through Abel's formula (5.3).

Definition. The Bessel function of the second kind of order zero, denoted by Y_0, is defined by, see (5.3):

$$Y_0(x) = J_0(x) \int \frac{\mathrm{d}x}{x\, J_0^2(x)}. \tag{7.9}$$

In this case, the formula (7.9) does not allow a simple symbolic approach; however, following [24], chapter 1, sect. 5, we can obtain the first terms of the power series, which reveal the logarithmic singularity in the origin of Y_0. From (7.3) we have, taking full advantage of computer algebra:

$$\frac{1}{xJ_0^2(x)} = \frac{1}{x} + \frac{x}{2} + \frac{5x^3}{32} + \frac{23x^5}{576} + \cdots$$

Hence, inserting in (7.9), we get, after integration

$$Y_0(x) = J_0(x) \left(\ln x + \frac{x^2}{4} + \frac{5x^4}{128} + \frac{23x^6}{3456} + \cdots \right) \tag{7.9b}$$

From (7.9b) we infer, recalling that $J_0(0) = 1$, that $Y_0(x) \to -\infty$ as $x \to 0^+$. From what has just been obtained, it follows that for each pair of constants c_1 and c_2, the function.

$$y(x) = c_1\, J_0(x) + c_2\, Y_0(x)$$

solves equation (7.2).

7.3 Bessel function of order one

We start observing that, since J_0 satisfies (7.2), differentiating we obtain:

$$xJ_0''' + 2J_0'' + xJ_0' + J_0 = 0\,. \tag{B$_0'$}$$

If we now express J_0 using (7.2) and then plug it in (B$_0'$) we obtain

$$xJ_0''' + J_0'' + (x^2 - 1)J_0' = 0.$$

This means that J_0' solves the Bessel equation of order 1. The Bessel function of order J_1 is conventionally defined as

$$J_1(x) = -J_0'(x). \tag{7.10}$$

Its transform is then obtained by applying the usual relations:

$$.\widehat{J_1}(s) = -\mathscr{L}\left[J_0'(x)\right](s) = -\left(s\widehat{J_0}(s) - J_0(0)\right) = 1 - \frac{s}{\sqrt{1+s^2}}.$$

Remark. Combining equations (7.10) with (7.6) the statement of the Theorem 7.2 can be rewritten in the following way

$$\int_0^x s\,J_0(\alpha\,s)\,J_0(\beta\,s)\,\mathrm{d}s = \frac{x}{\beta^2-\alpha^2}\Big(\beta^2\,J_1(\beta\,x)J_0(\alpha\,x) - \alpha^2\,J_1(\alpha\,x)J_0(\beta\,x)\Big). \tag{7.11}$$

Looking again at equation (7.11), we can observe that the particular situation $\alpha = \beta$ leads to an indeterminate form. The following result, also due to Lommel, deals with this problem and, to do so, uses, as in the case of the Theorem 7.2, the fact that $J_0(\alpha\,x)$ is a solution of the equation (7.7a).

Theorem. For any $x,\,\alpha \in \mathbb{R}$:

$$\int_0^x s\,J_0^2(\alpha\,s)\,\mathrm{d}s = \frac{x^2}{2}\Big(J_0^2(\alpha\,x) + J_1^2(\alpha\,x)\Big)\,. \tag{7.12}$$

Proof. For ease of writing, we repeat what we did in the proof of the Theorem 7.2 setting $u(x) = J_0(\alpha\,x)$ so that u solves the equation (7.7a). Now, after multiplying the two sides of (7.7a) by $2\,x\,u'$, we see, after some calculations, that this equation can be rewritten in the following form:

$$\frac{\mathrm{d}}{\mathrm{d}x}\left(x^2\,(u')^2 + \alpha^2x^2u^2\right) - 2\alpha^2xu^2 = 0.$$

Hence, after choosing $x \in \mathbb{R}$ integrating both sides in $[0, x]$ we see, also recalling that $u(x) = J_0(\alpha\,x)$ implies $u'(x) = \alpha\,J_0'(\alpha\,x) = -\alpha\,J_1(\alpha\,x)$, that

$$2\alpha^2 \int_0^x sJ_0^2(\alpha\,s)\mathrm{d}s = x^2\alpha^2J_1^2(\alpha\,x) + \alpha^2x^2J_0^2(\alpha\,x)$$

which leads to (7.12). □

Remark. The formula (7.12) can also be obtained, using de l'Hopital's Theorem, by taking the limit for $\beta \to \alpha$ in formula (7.6).

7.4 Bessel function of the first kind of any order

With the Frobenius method method, we can represent the solutions of the equation (7.1) using power series. It is almost impossible to list all the references covering this procedure; here, we limit ourselves to mentioning three of the best, in our opinion: [6], [27], and [36]. It is worth noting that it is also possible to follow the Laplace transform approach to this aim; in general, we highlight the classical treatment of [38].

When we identify the Bessel functions of the first kind as particular solutions of the equation (7.1), we intend to select solutions defined in the whole complex plane when $\nu > 0$. Consequently, the reader must be warned that our focus in this chapter is not on determining the equation's general solution but on determining functions with specific properties. Furthermore, as it is reasonable to foresee, the values assumed by the ν parameter influence the nature of the solutions of (7.1): the integer and semi-integer values are indeed peculiar. The cases $\nu = \pm 1/2$ in particular lead to the elementary solution of (7.1), in fact using the change of dependent variable $y = x^\nu u$ (7.1) is transformed into:

$$x\,u'' + (1+2\nu)u' + x\,u = 0\,. \tag{7.13}$$

Clearly for $\nu = -1/2$ (7.13) has general solution $u = c_1 \cos x + c_2 \sin x$ and equation (7.1) for $\nu = -1/2$ has the general solution

$$y = \frac{1}{\sqrt{x}}\left(c_1 \cos x + c_2 \sin x\right)\,. \tag{7.14}$$

The case $\nu = 1/2$ is analogous, here we use the variable transformation $y = x^{-\nu}u$, which leads to

$$x\,u'' + (1-2\nu)u' + x\,u = 0\,. \tag{7.15}$$

Solution via Frobenius method

Since we already examined the semi-integer case, we can assume $\nu \neq \pm 1/2$. Suppose that the solution $y(x)$ of the equation (7.1) has the form (5.22) presented at page 154: and its first two derivatives are given by (5.24), so inserting into (7.1) we get:

$$\sum_{n=0}^{\infty}\Big((\alpha+n)^2-\nu^2\Big)a_n x^{n+\alpha} + \sum_{n=2}^{\infty} a_{n-2}x^{n+\alpha} = 0.$$

Every coefficient of the powers of x has to be zero. Thus, we have the following conditions:

$$\left(\alpha^2-\nu^2\right)a_0 = 0 \tag{7.16a}$$

$$\left((\alpha+1)^2-\nu^2\right)a_1 = 0 \tag{7.16b}$$

$$\left((\alpha+n)^2-\nu^2\right)a_n + a_{n-2} = 0 \tag{7.16c}$$

If the parameter α is used to fulfil equation (7.16a), i.e. $\alpha = \pm\nu$, in (7.16b) must be $a_1 = 0$, i.e. the first derivative of the solution in origin vanishes. For this reason, in this situation, it makes sense to assume that $a_0 \neq 0$ to avoid trivial solutions, given that, in this case, we also have $a_1 = 0$. We begin with $\alpha = \nu$. The remaining equations henceforth become

$$(1+2\nu)\,a_1 = 0, \tag{7.17a}$$

$$n\,(2\nu+n)\,a_n + a_{n-2} = 0, \quad n \geq 2, \tag{7.17b}$$

where (7.17b) is obtained from (7.16c). Recalling our preliminary discussion, from the fact that $\nu \neq -1/2$ and assuming that 2ν is not a negative integer, from (7.17a) and (7.17b) follow:

$$a_0 \neq 0, \quad a_1 = 0, \qquad a_n = -\frac{a_{n-2}}{n(2\nu+n)}, \; n \geq 2.$$

This means that the coefficients of the odd index are all zero and that the following formula determines the ones with an even index:

$$a_{2n} = -\frac{a_{2n-2}}{2^2 n(\nu+n)}, \; n \in \mathbb{N}. \tag{7.18}$$

We can employ the "column"method starting from (7.18) obtaining:

$$\begin{aligned} a_2 &= -\frac{a_0}{2^2 \cdot 1 \cdot (1+\nu)} \\ a_4 &= -\frac{a_2}{2^2 \cdot 2 \cdot (2+\nu)} \\ a_6 &= -\frac{a_4}{2^2 \cdot 3 \cdot (3+\nu)} \\ a_8 &= -\frac{a_6}{2^2 \cdot 4 \cdot (4+\nu)} \\ &\cdots\cdots \end{aligned}$$

Henceforth, we can infer the general term using a Pochhamer symbol:

$$a_{2n} = \frac{(-1)^n}{2^{2n}} \frac{a_0}{n! \prod_{k=1}^{n}(k+\nu)} = a_{2n} = \frac{(-1)^n}{2^{2n}} \frac{a_0}{n!\,(1+\nu)_n}$$

Employing instead the gamma function, we have

$$(1+\nu)_n = \frac{\Gamma(1+\nu+n)}{\Gamma(1+\nu)}.$$

Therefore, we arrive at

$$a_{2n} = \frac{(-1)^n}{2^{2n}} \frac{\Gamma(1+\nu)}{n!\,\Gamma(1+\nu+n)} a_0.$$

Thus, the solution of (7.1) obtained in this way is given by

$$y = \sum_{n=0}^{\infty} a_{2n} x^{2n+\nu} = a_0 \sum_{n=0}^{\infty} \frac{(-1)^n}{2^{2n}} \frac{\Gamma(1+\nu)}{n!\,\Gamma(1+\nu+n)} x^{2n+\nu}.$$

The Bessel function of the first kind and order ν, which is denoted by $J_\nu(x)$, is indeed obtained by choosing:

$$a_0 = \frac{1}{2^\nu\,\Gamma(1+\nu)}.$$

Leading, if 2ν is not a negative integer, to

$$J_\nu(x) = \sum_{n=0}^{\infty} \frac{(-1)^n}{n!\,\Gamma(1+\nu+n)} \left(\frac{x}{2}\right)^{2n+\nu}. \tag{7.19}$$

A similar process can be followed when $\alpha = -\nu$. In this case, we get

$$J_{-\nu}(x) = \sum_{n=0}^{\infty} \frac{(-1)^n}{n!\,\Gamma(1-\nu+n)} \left(\frac{x}{2}\right)^{2n-\nu}. \tag{7.20}$$

This time, we must impose that 2ν is not a positive integer. Using the ratio test, it is possible to show that the series (7.19) and (7.20) are (absolutely) convergent for all $x \in \mathbb{R}$ (as a matter of fact also in $\mathbb{C}$).

Proposition 7.4. When ν is not an integer, functions J_ν and $J_{-\nu}$ are linearly independent: they form a basis for the vector space of the solutions of (7.1).

Proof. This fact follows immediately because both series (7.19) and (7.20) contain distinct powers of x, so they cannot be proportional, thus generating a non-zero Wronskian; therefore, they form a basis for the vector space of the solutions of (7.1). □

Indeed, it is possible, interesting and even instructive to obtain an explicit expression of the Wronskian $W(J_\nu, J_{-\nu})(x)$.

Proposition 7.5.

$$W(J_\nu, J_{-\nu})(x) = -\frac{2\sin(\nu\,\pi)}{\pi\,x}. \tag{7.21}$$

Proof. First, we observe that equation (5.2) here written as:

$$W(J_\nu, J_{-\nu})(x) = \frac{c}{x},$$

being c a constant to be detected. For this purpose, using the power series representations arrested at the first power:

$$J_\nu(x) = \frac{\left(\frac{x}{2}\right)^\nu}{\Gamma(1+\nu)}\,(1+\mathrm{o}(x)), \quad J_{-\nu}(x) = \frac{\left(\frac{x}{2}\right)^{-\nu}}{\Gamma(1-\nu)}\,(1+\mathrm{o}(x)) \tag{7.22}$$

we can evaluate the Wronskian for $x \to 0$ using the reflection formula (2.32) arriving at (7.21). □

Remark. From (7.22) we also infer that, if $\nu > 0$, for $x \to 0$ the function $J_\nu(x)$ vanishes while $J_{-\nu}(x)$ is unbounded.

Zeros

To study the nature of the zeros of the solutions of the equation (7.1), we exploit the Theorem 5.3 by comparing it with the differential equation $y'' + y = 0$, which characterises the trigonometric functions. For this purpose, we must first put the Bessel equation in normal form through the change of dependent variable $y = x^{1/2}u$, obtaining:

$$u'' + \left(1 + \frac{1-4\nu^2}{4x^2}\right) u = 0.$$

Hence we can infer thtat

- if $0 \le \nu < \frac{1}{2}$ then every interval of length π contains at least one zero of any solution of (7.1), since in this case $1 + \frac{1-4\nu^2}{4x^2} > 1$,
- if $\nu = \frac{1}{2}$ the distance between two zeros of any solution of (7.1) is exactly π (see also formula (7.14)),
- if $\nu > \frac{1}{2}$ then every interval of length π contains at most one zero of any solution of (7.1), since in this case $1 + \frac{1-4\nu^2}{4x^2} < 1$.

Recurrence properties

As the Bessel functions are the solutions of a family of differential equations depending on a parameter, we obtain several interesting recursive properties of these functions. As seen from the demonstrations we will provide, it should be emphasised that these properties hold for indices that are not necessarily integers.

Theorem 7.6. The following relations hold

$$(x^\nu J_\nu(x))' = x^\nu J_{\nu-1}(x) \tag{7.23a}$$

$$\left(x^{-\nu} J_\nu(x)\right)' = -x^{-\nu} J_{\nu+1}(x) \tag{7.23b}$$

$$J_\nu'(x) + \frac{\nu}{x} J_\nu(x) = J_{\nu-1}(x) \tag{7.23c}$$

$$J_\nu'(x) - \frac{\nu}{x} J_\nu(x) = -J_{\nu+1}(x). \tag{7.23d}$$

Proof. To show (7.23a) multiply the series (7.19) by x^ν and then differentiate term by term with respect to x, obtaining:

$$(x^\nu J_\nu(x))' = \left(\sum_{n=0}^{\infty} \frac{(-1)^n x^{2n+2\nu}}{2^{2n+\nu} n!\,\Gamma(1+\nu+n)}\right)' = \sum_{n=0}^{\infty} \frac{(-1)^n (2n+2\nu) x^{2n+2\nu-1}}{2^{2n+\nu} n!\,\Gamma(1+\nu+n)}$$

$$= \sum_{n=0}^{\infty} \frac{(-1)^n\, 2(n+\nu) x^{2n+2\nu-1}}{2^{2n+\nu} n!\,(\nu+n)\Gamma(\nu+n)} = \sum_{n=0}^{\infty} \frac{(-1)^n\, x^{2n+2\nu-1}}{2^{2n+\nu-1} n!\,\Gamma(\nu+n)}$$

$$= x^\nu \sum_{n=0}^{\infty} \frac{(-1)^n}{n!\,\Gamma(\nu+n)} \left(\frac{x}{2}\right)^{2n+\nu-1} = x^\nu J_{\nu-1}(x)\,.$$

With very similar reasoning, we obtain (7.23b). Eventually, calculating the two derivatives on the left-hand side of (7.23a) and (7.23b), we find (7.23c) and (7.23d). □

Remark.

- Looking at (7.23d) if we take $\nu = 0$, we confirm the convention established formerly in equation (7.10).
- If we subtract the two sides of (7.23c) and (7.23d), we obtain a recursive relation that does not use derivatives:

$$J_\nu(x) = \frac{x}{2\nu}\Big(J_{\nu-1}(x) + J_{\nu+1}(x)\Big). \tag{7.24}$$

7.5 Bessel functions of order $\pm\frac{1}{2}$

We continue our analysis on cases involving semi-integer orders with $\nu = \pm 1/2$. These cases are connected to equation (7.14). However, these formulas represent the general solutions of the Bessel equation of order $\pm 1/2$. Our focus is on the two solutions identified by the power series (7.19) and (7.20), which, given the aforementioned (7.14), are elementary functions. The interest in the case of order $\pm 1/2$ depends both on various applications and on the fact that the Bessel functions of these orders and all orders of the form $\nu = \pm 1/2 + n$ with $n \in \mathbb{Z}$ characterise occurrences of the Riccati equation, see Section 8.3, which can be solved in terms of elementary functions.

Theorem 7.7. The expression for $J_{1/2}$ and $J_{-1/2}$ are respectively:

$$J_{1/2}(x) = \sqrt{\frac{2}{\pi x}}\,\sin x \tag{7.25a}$$

$$J_{-1/2}(x) = \sqrt{\frac{2}{\pi x}}\,\cos x. \tag{7.25b}$$

Proof. In order to compute $J_{1/2}(x)$ we have to sum, remembering (7.19), a series of the form

$$J_{1/2}(x) = \left(\frac{x}{2}\right)^{1/2} \sum_{n=0}^{\infty} \frac{(-1)^n x^{2n}}{2^{2n} n!\,\Gamma(\frac{3}{2} + n)}. \tag{7.19a}$$

We need to combine two formulas from chapter 2 to evaluate the Gamma term in (7.19a), namely (2.23) and (2.4) in order to arrive at, if $n \in \mathbb{N}$:

$$\Gamma\left(\tfrac{3}{2} + n\right) = \frac{\sqrt{\pi}(2n+1)(2n)!}{2^{2n+1} n!}. \tag{2.23a}$$

Therefore plugging (2.23a) into (7.19a) we obtain, after some adjustments:

$$J_{1/2}(x) = \left(\frac{x}{2}\right)^{1/2} \sum_{n=0}^{\infty} \frac{(-1)^n 2 x^{2n}}{(2n+1)!\sqrt{\pi}} = \sqrt{\frac{2x}{\pi}}\,\frac{1}{x} \sum_{n=0}^{\infty} \frac{(-1)^n x^{2n+1}}{(2n+1)!}.$$

The latter, after simplification, leads to (7.25a). The way to arrive at (7.25b) is similar, starting from (7.20), so we leave the details to the Reader. □

Remark 7.8. Combining Theorem 7.7 with the recurrence relation (7.24), we acknowledge that the Bessell functions of semi-integer order, i.e. order $\nu = \pm 1/2 + n$ with $n \in \mathbb{Z}$, are elementary. For instance, from (7.24) we see that

$$J_{3/2}(x) = \frac{1}{x} J_{1/2}(x) - J_{-1/2}(x)$$

which from (7.25a) and (7.25b) gives

$$J_{3/2}(x) = \sqrt{\frac{2}{\pi x}} \left(\frac{\sin x}{x} - \cos x \right) . \tag{7.25c}$$

7.6 Bessel functions of integer order

Formula (7.21) shows that for $\nu = m \in \mathbb{N}$, the two functions J_m and J_{-m} are not independent. This fact can also be understood by reasoning directly about power series; in fact, in the series that defines J_m, one of the denominator terms would have a starting value where the reciprocal gamma function has an integer argument and, therefore, vanishes; in other terms, we have that in this particular situation (7.20) becomes:

$$J_{-m}(x) = \sum_{n=m}^{\infty} \frac{(-1)^n}{n!\, \Gamma(1-m+n)} \left(\frac{x}{2} \right)^{2n-m} . \tag{7.20'}$$

At this point, it is convenient to rescale the summation index n to start from zero, i.e. $j = n - m$ so that (7.20') becomes

$$\begin{aligned} J_{-m}(x) &= \sum_{j=0}^{\infty} \frac{(-1)^{j+m}}{(j+m)!\, \Gamma(1-m+j+m)} \left(\frac{x}{2} \right)^{2(j+m)-m} \\ &= (-1)^m \sum_{j=0}^{\infty} \frac{(-1)^j}{(j+m)!\, j!} \left(\frac{x}{2} \right)^{2j+m} \\ &= (-1)^m J_m(x) . \end{aligned} \tag{7.26}$$

From this discussion, it is clear that to deal with Bessel functions of integer order, we must first complete the general solution of the differential equation (7.1) by introducing the so-called Bessel functions of the second kind. We will discuss these problems in the next chapter 8.

7.7 Generating function

As in the case of the Hermite functions, Bessel functions can be introduced through a suitable generating function. Even if this approach is more axiomatic than genetic, it allows us to establish properties of considerable interest quickly. Let us consider the function of two variables:

$$g(x,t) = \exp\left[\frac{x}{2}\left(t - \frac{1}{t}\right)\right]. \tag{7.27}$$

Theorem 7.9. Bessel functions of integer order $J_m(x)$, $m \in \mathbb{Z}$ can be defined expanding in power series with respect to t the generating function (7.27) as follows:

$$g(x,t) = \sum_{m=-\infty}^{+\infty} J_m(x)\, t^m. \tag{7.28}$$

Proof. The $J_m(x)$'s are obtained through the product of the two Maclaurin series, one in x, and the other in t, generated by the decomposition of (7.27) into two exponential factors:

$$\begin{aligned} g(x,t) &= \exp\left(\frac{xt}{2}\right)\exp\left(-\frac{x}{2t}\right) = \sum_{j=0}^{\infty}\frac{(tx)^j}{2^j\, j!}\sum_{k=0}^{\infty}\frac{(-x)^k}{(2t)^k\, k!} \\ &= \sum_{j=0}^{\infty}\sum_{k=0}^{\infty}(-1)^k\left(\frac{x}{2}\right)^{j+k}\frac{t^{j-k}}{j!\,k!}. \end{aligned}$$

At this point, to obtain a power series in t alone, we use the index change $j - k = m$ so that, since both j and k have a range between all nonnegative integers, it follows that m spans $\mathbb{Z}$ and we can rewrite the last sum as follows:

$$g(x,t) = \sum_{m=-\infty}^{\infty}\sum_{k=0}^{\infty}(-1)^k\left(\frac{x}{2}\right)^{2k+m}\frac{t^m}{k!(k+m)!}.$$

In the series indexed by k, we recognise the Bessel function of integer index $J_m(x)$ and, therefore, (7.28) follows. □

Remark. Recalling (7.26), we can express (7.28) using nonnegative indexes only:

$$g(x,t) = J_0(x) + \sum_{m=0}^{+\infty} J_m(x)\ \left(t^m + (-1)^m t^m\right). \tag{7.29}$$

Remark. Following the exponential structure of the first argument of the generating function, we have that $g(x+y,t) = g(x,t)\,g(y,t)$. Therefore, recalling the Cauchy product between two power series, we have that

$$J_m(x+y) = \sum_{k=-\infty}^{+\infty} J_k(x)\, J_{m-k}(y). \tag{7.30}$$

7.8 First integral representation

Using the representation of Bessel functions based on the generating function, we obtain their integral representation for non-negative integer orders, generalising and, more importantly, motivating formula (7.4). We follow the argument presented in [27].

Theorem 7.10. For any $m \in \mathbb{Z}$, $m \geq 0$:

$$J_m(x) = \frac{1}{\pi} \int_0^{\pi} \cos(m\,\vartheta - x \sin\vartheta)\, \mathrm{d}\vartheta\,. \tag{7.31}$$

Proof. In (7.27) choose $t = e^{i\,\vartheta}$. Then, since $t - \frac{1}{t} = e^{i\,\vartheta} - e^{-i\,\vartheta} = 2\,i\,\sin\vartheta$, the left-hand side of (7.29), recalling (7.27), reads as $e^{ix\sin\vartheta}$. Moreover, for de Moivre formula we have

$$t^m = \cos(m\vartheta) + i\,\sin(m\vartheta), \quad t^{-m} = \cos(m\vartheta) - i\,\sin(m\vartheta)\,.$$

Summarizing (7.29) this situation becomes

$$\begin{aligned} e^{ix\sin\vartheta} &= J_0(x) + \sum_{m=1}^{\infty} (1 + (-1)^m)\, J_m(x) \cos(m\vartheta) + \\ &+ i \sum_{m=1}^{\infty} (1 - (-1)^m)\, J_m(x) \sin(m\vartheta). \end{aligned} \tag{7.32}$$

From this relation, equating the real and imaginary parts of both sides separately, we obtain

$$\cos(x\sin\vartheta) = J_0(x) + \sum_{m=1}^{\infty} (1 + (-1)^m)\, J_m(x)\cos(m\vartheta)\,. \tag{7.33a}$$

$$\sin(x\sin\vartheta) = \sum_{m=1}^{\infty} (1 - (-1)^m)\, J_m(x)\sin(m\vartheta)\,. \tag{7.33b}$$

Let's fix $x \in \mathbb{R}$ in the two formulas (7.33a) and (7.33b). We recognise the Fourier series (sine and cosine) of the two functions $c(\varphi) = \cos(x\sin\varphi)$ and $s(\vartheta) = \sin(x\sin\vartheta)$ where $0 < \vartheta < \pi$. Hence, we infer that

$$(1 + (-1)^m)\, J_m(x) = \frac{2}{\pi}\int_0^{\pi} \cos(x\sin\vartheta)\cos(m\vartheta)\,\mathrm{d}\vartheta\,. \tag{7.34a}$$

$$(1 - (-1)^m)\, J_m(x) = \frac{2}{\pi}\int_0^{\pi} \sin(x\sin\vartheta)\sin(m\vartheta)\,\mathrm{d}\vartheta\,. \tag{7.34b}$$

Notice that (7.34a) holds for $n \in \mathbb{Z}$, $n \geq 0$ and (7.34b) is valid for $n \geq 1$. Hence adding (7.34a) and (7.34b) we obtain

$$2J_m(x) = \frac{2}{\pi}\int_0^{\pi} \Big(\cos(m\vartheta)\cos(x\sin\vartheta) + \sin(m\vartheta)\sin(x\sin\vartheta)\Big)\,\mathrm{d}\vartheta\,.$$

The formula (7.31) follows from the addition formula for the cosine. □

Remark. Fourier series (7.32) can be written using complex exponential as

$$e^{ix \sin\vartheta} = \sum_{m=-\infty}^{\infty} e^{im\vartheta} J_m(x) .$$

Hence using Parseval identity (1.28), we infer that

$$\sum_{m=-\infty}^{\infty} J_m^2(x) = 1.$$

Remark 7.11. Theorem 7.10 implies that

$$|J_m(x)| \le 1 \text{ and } \left|J_m^{(k)}(x)\right| \le 1,\, k \ge 1 .$$

7.9 Second integral representation

In this section, we present an alternative integral representation for the function of the first kind due to Bessel himself, aiming to extend such representation to non-integer indexes; we recall that Theorem 7.10 is based on the representation in Fourier series of the function $f(x) = e^{ix\sin\vartheta}$, and this implies that the representation formula obtained applies only to functions of integer order. Since getting to the two representation formulas takes considerable work, we divide it into several steps. We, therefore, begin with a preliminary lemma.

Lemma 7.12. If $\nu > -1/2$ the differential equation:

$$x z'' + (1 + 2\nu)\, z' + x z = 0 \tag{7.35}$$

is solved by:

$$z_\nu(x) := \int_0^1 \left(1 - t^2\right)^{\nu-\frac{1}{2}} \cos(t\, x)\, \mathrm{d}t. \tag{7.36}$$

Proof. The conditions for deriving under the sign of integral are satisfied since we have chosen $\nu > -1/2$, so we have:

$$z_\nu'(x) = \int_0^1 -t \left(1 - t^2\right)^{\nu-\frac{1}{2}} \sin(t\, x)\, \mathrm{d}t\,, \tag{7.37a}$$

$$z_\nu''(x) = \int_0^1 -t^2 \left(1 - t^2\right)^{\nu-\frac{1}{2}} \cos(t\, x)\, \mathrm{d}t\,. \tag{7.37b}$$

Integrating by parts in (7.37a), we get

$$z_\nu'(x) = -x \int_0^1 \frac{\left(1-t^2\right)^{\nu+\frac{1}{2}}}{2\nu+1} \cos(t\,x)\,\mathrm{d}t\,. \tag{7.37aa}$$

Thus inserting (7.36), (7.37aa) and (7.37b) on the left-hand side of (7.35) we obtain

$$x \int_0^1 \cos(t\,x) \left(-t^2\left(1-t^2\right)^{\nu-\frac{1}{2}} - \left(1-t^2\right)^{\nu+\frac{1}{2}} + \left(1-t^2\right)^{\nu-\frac{1}{2}}\right) \mathrm{d}t.$$

Verifying that the quantity in the parenthesis vanishes is a straightforward computation, proving (7.35). □

We use lemma 7.12 to prove a second useful result.

Lemma 7.13. For $\nu > -\frac{1}{2}$ and $x > 0$, the parametric integral

$$F_\nu(x) = x^\nu \int_0^1 \left(1-t^2\right)^{\nu-\frac{1}{2}} \cos(t\,x)\,\mathrm{d}t \tag{7.38}$$

solves the Bessel equation (7.1).

Proof. Let's change the dependent variable in (7.1) as $y = x^\nu\, z$ so that $y' = x^{\nu-1}\,(xz' + \nu z)$ and

$$y'' = x^\nu \left(\frac{(\nu-1)\nu z}{x^2} + z'' + \frac{2\nu z'}{x}\right).$$

Inserting in (7.1), we obtain, after some algebraic work

$$x^{\nu+1}\,(xz'' + (2\nu+1)z' + xz) = 0\,. \tag{7.39}$$

From lemma 7.12, equation (7.35) we know that $z_\nu(x)$ given by (7.36) solves (7.39), thus $y(x) = x^\nu\, z_\nu(x)$ solves (7.1) ending the proof. □

We can state the second integral representation theorem for J_ν. The procedure we present, inspired by [61], Chapter 3, is used to prove that the function F_ν obtained in (7.38) is, in fact, proportional to the power series (7.19).

Theorem 7.14. For $\nu > -\frac{1}{2}$ and $x > 0$, the following integral representations for the Bessel function of the first kind hold:

$$J_\nu(x) = \frac{2\left(\frac{x}{2}\right)^\nu}{\sqrt{\pi}\,\Gamma(\nu+\frac{1}{2})} \int_0^{\frac{\pi}{2}} \cos(x\,\sin u)\,\cos^{2\nu} u\,\mathrm{d}u\,, \tag{7.40a}$$

$$J_\nu(x) = \frac{\left(\frac{x}{2}\right)^\nu}{\sqrt{\pi}\,\Gamma(\nu+\frac{1}{2})} \int_0^{\pi} \cos(x\,\sin u)\,\cos^{2\nu} u\,\mathrm{d}u\,, \tag{7.40b}$$

$$J_\nu(x) = \frac{2\left(\frac{x}{2}\right)^\nu}{\sqrt{\pi}\,\Gamma(\nu+\frac{1}{2})} \int_0^{\frac{\pi}{2}} \cos(x\,\cos u)\,\sin^{2\nu} u\,\mathrm{d}u\,, \tag{7.40c}$$

$$J_\nu(x) = \frac{\left(\frac{x}{2}\right)^\nu}{\sqrt{\pi}\,\Gamma(\nu+\frac{1}{2})} \int_0^{\pi} \cos(x\,\cos u)\,\sin^{2\nu} u\,\mathrm{d}u\,. \tag{7.40d}$$

Proof. By the lemma 7.12, we know that F_ν solves the Bessel equation (7.1): to complete the proof, we must show that this function is proportional to the function J_ν introduced in (7.19). We start using the change of variable $t = \sin u$ in (7.38) we can express F_ν as:

$$F_\nu(x) = x^\nu \int_0^{\frac{\pi}{2}} \cos(x \sin u) \cos^{2\nu} u \, du. \tag{7.38a}$$

Applying the power series representation of the cosine, see page 10, to $\cos(x \sin u)$, we can express the integrand on the right-hand side of (7.38a) as:

$$\cos(x \sin u) \cos^{2\nu} u = \sum_{n=0}^{\infty} \frac{(-1)^n}{(2n)!} x^{2n} \sin^{2n} u \cos^{2\nu} u \, .$$

In this way, we can integrate by series, obtaining:

$$F_\nu(x) = \int_0^{\frac{\pi}{2}} \cos(x \sin u) \cos^{2\nu} u \, du = \sum_{n=0}^{\infty} \frac{(-1)^n}{(2n)!} x^{2n} \mathcal{I}_n \, ,$$

where, using (2.25a) and Theorem 2.9, see page 68:

$$\mathcal{I}_n = \int_0^{\frac{\pi}{2}} \sin^{2n} u \cos^{2\nu} u \, du = \frac{\Gamma\left(\nu + \frac{1}{2}\right) \Gamma\left(n + \frac{1}{2}\right)}{2\Gamma(n + \nu + 1)} \, .$$

Now, since n is a positive integer, we can invoke again identity (2.23), whence

$$F_\nu(x) = \frac{\sqrt{\pi}}{2} \Gamma\left(\nu + \frac{1}{2}\right) \sum_{n=0}^{\infty} (-1)^n \frac{x^{2n}}{2^{2n} \, n! \, \Gamma(n + \nu + 1)} .$$

Now we indeed recognise that $F_\nu(x)$ and $J_\nu(x)$ are proportional, thus using an appropriate arrangement of a multiplicative constant in the last expression for F_ν we arrive at (7.40a). The remaining formulas (7.40b), (7.40c), and (7.40d) follow from symmetry properties of the integral, [100, page 266 Corollary 10.32]: in fact, recalling that when $f(\pi - u) = f(u)$, then

$$\int_0^{\pi} f(u) du = 2 \int_0^{\frac{\pi}{2}} f(u) du \, .$$

(7.40b) follows since $f(u) = \cos(x \sin u) \cos^{2\nu} u = f(\pi - u)$. Then, for the fact that:

$$\int_0^{\pi} f(u) du = \int_0^{\pi} f(\pi - u) du \, ,$$

we see that (7.40b) implies (7.40d) and finally, again by symmetry, (7.40d) implies (7.40c) ending the proof of Theorem 7.14. □

We exploit the integral representation (7.40a) to obtain Weber's integral formula, inspired by the approach proposed in [6]. We have chosen to present the demonstration in great detail, as it is very interesting since three essential properties of the gamma function are necessary. This makes us realise how close the correlations between diverse families of special functions are

Theorem. If $0 < m < 1/2$ and $\nu > -1/2$ then

$$\int_0^\infty x^{2m-\nu-1} J_\nu(x)\,\mathrm{d}x = \frac{2^{2m-p-1}\Gamma(m)}{\Gamma(\nu - m - 1)}\,. \tag{7.41}$$

Proof. For simplicity, we write

$$W_{m,\nu} = \int_0^\infty x^{2m-\nu-1} J_\nu(x)\,\mathrm{d}x.$$

Then, rewriting the left-hand side of (7.41) using the integral representation (7.40a) of $J_\nu(x)$, notice that we took the constant out of the integrals in the process:

$$W_{m,\nu} = \frac{2^{1-\nu}}{\sqrt{\pi}\,\Gamma\left(\nu + \frac{1}{2}\right)} \int_0^\infty x^{2m-1} \left(\int_0^{\frac{\pi}{2}} \cos(x \sin u)\, \cos^{2\nu} u\,\mathrm{d}u \right) \mathrm{d}x\,. \tag{7.42}$$

Since the exchange of the order of integration is permissible, we have the following:

$$W_{m,\nu} = \frac{2^{1-\nu}}{\sqrt{\pi}\,\Gamma\left(\nu + \frac{1}{2}\right)} \int_0^{\frac{\pi}{2}} \cos^{2\nu} u \left(\int_0^{+\infty} x^{2m-1} \cos(x \sin u)\,\mathrm{d}x \right) \mathrm{d}u\,. \tag{7.42a}$$

To evaluate the inner integral in (7.42a) we refer to equation (2.8aa), see page 61, which provides:

$$W_{m,\nu} = \frac{2^{1-\nu} \cos(m\pi)\,\Gamma(2m)}{\sqrt{\pi}\,\Gamma\left(\nu + \frac{1}{2}\right)} \int_0^{\frac{\pi}{2}} \cos^{2\nu} u\, \sin^{-2m} u\,\mathrm{d}u. \tag{7.42b}$$

The integral on the right-hand side of (7.42b) is promptly evaluated in terms of the beta function, accordingly to (2.25a) so that

$$W_{m,\nu} = \frac{2^{1-\nu} \cos(m\pi)\,\Gamma(2m)}{\sqrt{\pi}\,\Gamma\left(\nu + \frac{1}{2}\right)} \frac{\Gamma(\frac{1}{2} - m)\Gamma(\frac{1}{2} + \nu)}{2\Gamma(1 + \nu - m)}\,. \tag{7.42c}$$

Eventually, the thesis follows from the relevant simplifications, replacing $\Gamma(2m)$ using Legendre duplication formula, Theorem 2.15, formula (2.50), and from the cosine duplication formula (2.34). □

Chapter 8

Bessel functions: Part 2

In the previous chapter, we chose not to deal primarily with the problem of finding the general solution of the Bessel equation (7.1), specifically in the integer coefficient case, focusing on the study of its particular solution J_ν introduced in (7.19) for $\nu > 0$ and in (7.20) for $\nu < 0$. In this chapter, to complete the description of the general solution of (7.1), we will introduce the Bessel functions of the second kind. We will also consider equations related to (7.1), as proposed in [24] with particular reference to the Airy equation (6.1) and some generalisation of it, and to the Riccati equation in its original form, which, led Daniel Bernoulli to consider problems related to the Bessel differential equation.

8.1 Bessel function of the second kind

In proposition 7.4, we detected the general solution of (7.1) for noninteger indexes. To obtain the general solution of the Bessel equation (7.1) when $\nu = n \in \mathbb{N}$, we need to introduce the Bessel functions of the second kind, which is defined as follows:

$$Y_\nu(x) = \frac{\cos(\nu\,\pi)J_\nu(x) - J_{-\nu}(x)}{\sin(\nu\,x)}. \tag{8.1}$$

The function introduced in (8.1) is, of course, a solution to (7.1) if we assume ν is not an integer. Furthermore, observe that for $\nu \to m \in \mathbb{N}$ the expression in (8.1) is indeterminate; hence for $m \in \mathbb{N}$ we define:

$$Y_m(x) = \lim_{\nu \to m} \frac{\cos(\nu\,\pi)J_\nu(x) - J_{-\nu}(x)}{\sin(\nu\,x)}.$$

We do not present here the computation of this limit, referring to [12, page 668] or [36, page 248] for details; here, we limit to reporting the result of the

DOI: 10.1201/9781003171546-8

computation, which is obtained using the de l'Hopital rule:

$$Y_m(x) = \frac{2}{\pi} J_m(x) \ln\left(\frac{x}{2}\right) - \frac{1}{\pi}\sum_{j=0}^{m-1} \frac{(m-j-1)!}{j!}\left(\frac{x}{2}\right)^{2j-m}$$
$$- \frac{1}{\pi}\sum_{j=0}^{\infty} \frac{(-1)^j}{j!(m+j)!}\left[\psi(j+1) + \psi(m+j+1)\right]\left(\frac{x}{2}\right)^{2j+m}.$$

Here, ψ is the digamma function discussed in chapter 3. Finally, observe that if $m \in \mathbb{N}$ then from (7.19) we find again that $J_m(x) = (-1)^m J_m(-x)$, this, in particular, implies that the Bessel functions of even integer index are even functions. In contrast, those of odd integer indexes are odd functions.

Knowing the general solution of the Bessel equation (7.1) is useful for dealing with linear equations other than (7.1) in the cases where we can determine an appropriate transformation of variables, as in the following example.

Example. Consider the differential equation:

$$y'' + \frac{1}{x^3}\, y = 0. \tag{8.2}$$

Note that for (8.2), we cannot exploit Frobenius's method since the singularity is of the third order. Changing the independent variable to:

$$x = \frac{4}{\xi^2} \iff \xi = \frac{2}{\sqrt{x}};$$

using (5.18) we see that the transformed equation (5.19) in $z(\xi) = y(2/\sqrt{x})$ in this case becomes

$$z'' + \frac{3}{\xi} z' + z = 0.$$

Finally replacing $z(\xi)$ with $u(\xi)/\xi$ we arrive at the u differential equation

$$\xi^2 u'' + \xi u' + \left(\xi^2 - 1\right) u = 0$$

which indeed is (7.1) with $\nu = 1$, hence

$$u(\xi) = k_1\, J_1(\xi) + k_2\, Y_1(\xi).$$

Going back, we find

$$z(\xi) = \frac{1}{\xi}\left(k_1\, J_1(\xi) + k_2\, Y_1(\xi)\right).$$

In conclusion, the general solution to (8.2) is

$$y = \sqrt{x}\left(k_1\, J_1\left(\tfrac{2}{\sqrt{x}}\right) + k_2\, Y_1\left(\tfrac{2}{\sqrt{x}}\right)\right).$$

8.2 Bessel related equations

The solution of equation (8.2) just discussed shows us how it is possible to try to use Bessel functions to solve larger classes of differential equations. On the other hand, the solution of the equation (8.2), brilliant as it is, makes us realise that for application purposes, it is essential to obtain a classification that allows us to quickly understand when a certain differential equation is treatable using Bessel functions. As we shall see, the analogy between trigonometric and hyperbolic functions is very close, in the sense that starting from the basic differential equations $y'' \pm a^2 y = 0$, we can use these functions to solve all the constant coefficients differential equations. In this section, we will deal with a class of differential equations related to (7.1), presenting the idea due to Bowman's following [24] and [6]. For $\nu \geq 0$, $c > 0$ and $a,\, b \neq 0$, consider the equation:

$$x^2 y'' + (1-2a)xy' + \left[b^2c^2x^{2c} + (a^2 - c^2\nu^2)\right] y = 0. \tag{8.3}$$

Theorem 8.1. The general solution to (8.3) is given by

$$y = x^a \Big(k_1\, J_\nu(b\,x^c) + k_2\, Y_\nu(b\,x^c)\Big), \tag{8.4}$$

being $k_1,\, k_2 \in \mathbb{R}$. Moreover, if ν is not an integer, the solution to (8.3) can be represented using Bessel functions of the first kind only, namely:

$$y = x^a \Big(k_1\, J_\nu(b\,x^c) + k_2\, J_{-\nu}(b\,x^c)\Big). \tag{8.4a}$$

Proof. We begin changing the dependent variable as $y = x^a\, z$ transforming (8.3) in

$$x^2\, z'' + x\, z' + (b^2c^2x^{2c} - c\nu^2)\, z = 0. \tag{8.3a}$$

Now we change the independent variable in (8.3a) as $t = x^c$ arriving, thanks to (5.19), at

$$t^2 z'' + tz' + (b^2t^2 - \nu^2)\, z = 0. \tag{8.3b}$$

The ultimate change of the independent variable is, at this point, straightforward: $t\,b = s$ which yields

$$s^2 z'' + sz' + (t^2 - \nu^2)z = 0. \tag{8.3c}$$

We have thus obtained a Bessel equation: once solved, we can proceed backwards to obtain the solution of the assigned equation (8.3):

- Solution to (8.3c) is $z(s) = k_1 J_\nu(s) + k_2 Y_\nu(s)$
- Solution to (8.3b) is $z(t) = k_1 J_\nu(b\,t) + k_2 Y_\nu(b\,t)$
- Solution to (8.3a) is $z(x) = k_1 J_\nu(b\,x^c) + k_2 Y_\nu(b\,x^c)$

We then can conclude that the solution to (8.3) is, indeed, (8.4).

□

Remark 8.2. Taking the particular case of $a = 0$, $c = 1$ in (8.3) we have that general solution of the equation:

$$x^2\, y'' + x\, y' + \left(b^2\, x^2 - \nu^2\right)\, y = 0 \tag{8.5}$$

is

$$y = k_1\, J_\nu(b\, x) + k_2\, Y_\nu(b\, x). \tag{8.6}$$

It is interesting to consider carefully the analogies with lemma 7.1.

Remark 8.3. Given the Airy-related equation

$$y'' + x\, y = 0. \tag{8.7}$$

In order to use equation (8.3) and its solution formula, we must multiply by x^2 obtaining $x^2 y'' + x^3\, y = 0$, so that we find:

$$a = \frac{1}{2}, \quad b = \frac{2}{3}, \quad c = \frac{3}{2}, \quad \nu = \frac{1}{3}.$$

Hence the general solution to (8.7) is

$$y = \sqrt{x}\, \left(c_1 J_{\frac{1}{3}}(\tfrac{2}{3}\, x^{3/2}) + c_2 Y_{\frac{1}{3}}(\tfrac{2}{3}\, x^{3/2}) \right).$$

Observe that here, since $\nu = \frac{1}{3} \notin \mathbb{Z}$, we can also express the found solution as:

$$y = \sqrt{x}\, \left(c_1 J_{\frac{1}{3}}(\tfrac{2}{3}\, x^{3/2}) + c_2 J_{-\frac{1}{3}}(\tfrac{2}{3}\, x^{3/2}) \right).$$

8.3 Connection with the Riccati equation

We consider here the particular Riccati equation (indeed the first kind of Riccati equation that was studied, dating back to the 18th century);

$$y' = \alpha\, x^p + \beta\, y^2. \tag{8.8}$$

Where α, β and p are real numbers. Using the variable transformation, due to Euler:

$$y = -\frac{u'}{\beta\, u}. \tag{8.9}$$

Equation (8.8) is transformed into an (Airy-like) linear equation of second order:

$$u'' + \alpha\beta\, x^p u = 0. \tag{8.10}$$

To solve (8.10) we use Theorem 8.1 after recognizing that, in this case, we have

$$a = \frac{1}{2}, \; c = \frac{2+p}{2}, \; b = \frac{2\sqrt{a\,b}}{2+p}, \; \nu = \frac{1}{2+p},$$

having assumed implicitly, for ease of computation, that $p + 2 > 0$. Hence

$$u(x) = \sqrt{x}\left(k_1 J_{\frac{1}{2+p}}\left(\tfrac{2\sqrt{a\,b}}{2+p}\,x^{\frac{2+p}{2}}\right) + k_2 Y_{\frac{1}{2+p}}\left(\tfrac{2\sqrt{a\,b}}{2+p}\,x^{\frac{2+p}{2}}\right)\right).$$

Thus, the solution to Riccati equation (8.8) is obtained via transformation (8.9).

It is remarkable that despite the simplicity of the structure of the equation (8.8), the solution is instead highly complicated. The first contributions to this equation, due among others to Riccati and Daniel Bernoulli, investigated the research of which exponents p of x in (8.8) originated elementary solutions; Daniel Bernoulli was able to determine the sequence of exponents corresponding to elementary solutions. We know semi-integer indices produce such elementary solutions in light of Remark 7.8. For example, if

$$\frac{1}{2+p} = \frac{3}{2} \iff p = -\frac{4}{3},$$

then

$$y = \frac{3\alpha \sin\left(3\sqrt{\alpha\,\beta}\,\sqrt[3]{x}\right)}{\sqrt[3]{x}\left(3\sqrt{\alpha\,\beta}\,\sqrt[3]{x}\cos\left(3\sqrt{\alpha\,\beta}\,\sqrt[3]{x}\right) - \sin\left(3\sqrt{\alpha\,\beta}\,\sqrt[3]{x}\right)\right)}$$

solves

$$y' = \alpha\, x^{-\frac{4}{3}} + \beta\, y^2.$$

The result established by Daniel Bernoulli [20] states that solutions of (8.8) are represented through elementary functions if the following condition holds:

$$p = -\frac{4n}{1+2n}, \qquad n \in \mathbb{Z}.$$

Later, Liouville proved in [76] that, beyond Bernoulli's conditions, equation (8.8) cannot be solved in terms of elementary functions.

8.4 Modified Bessel functions of the first kind

To motivate the reasons why it is useful to introduce modified Bessel functions, we find it useful to start with an example that is interesting in itself. Consider again, in view of Theorem 8.1 the Airy equation:

$$y'' - x\,y = 0 \tag{6.1}$$

we already studied in chapter 6. If we multiply by x^2, obtaining $x^2\,y'' - x^3\,y = 0$, referring to (8.3) we see that $a = \frac{1}{2}, \quad b = \frac{2}{3}\,i, \quad c = \frac{3}{2}, \quad \nu = \frac{1}{3}$, leading to the solution

$$y = \sqrt{x}\left(k_1 J_{\frac{1}{3}}\left(\tfrac{2}{3}\,i\,x^{3/2}\right) + k_2 Y_{\frac{1}{3}}\left(\tfrac{2}{3}\,i\,x^{3/2}\right)\right) \tag{8.11}$$

or, since $\nu = \frac{1}{3} \notin \mathbb{Z}$:

$$y = \sqrt{x}\left(k_1 J_{\frac{1}{3}}\left(\tfrac{2}{3}\,i\,x^{3/2}\right) + k_2 J_{-\frac{1}{3}}\left(\tfrac{2}{3}\,i\,x^{3/2}\right)\right). \tag{8.11a}$$

A similar situation occurs by slightly changing the equation (8.5) discussed in the remark 8.2 to

$$x^2 y'' + x y' - \left(b^2 x^2 + \nu^2\right) y = 0\,. \tag{8.5a}$$

Since the parameter now is $i\,b$ solution to (8.11a) is

$$y = k_1 J_\nu(i\,b\,x) + k_2 Y_\nu(i\,b\,x)\,. \tag{8.6a}$$

In applications, it would be preferable to avoid the appearance of complex arguments. For this reason, it is necessary to use a procedure similar to that used in the elementary case, leading to trigonometric functions starting from the differential equation $y'' + a^2\,y = 0$. We then restart from the power series (7.19), for $\nu > 0$ in which the argument is changed to $i\,x$

$$J_\nu(i\,x) = \sum_{n=0}^{\infty} \frac{(-1)^n}{n!\,\Gamma(1+\nu+n)} \left(\frac{i\,x}{2}\right)^{2n+\nu}. \tag{8.12}$$

Therefore, by carrying the imaginary unit out of the summation, we get $i^{2n} = (i^2)^n = 1$:

$$J_\nu(i\,x) = i^\nu \sum_{n=0}^{\infty} \frac{1}{n!\,\Gamma(1+\nu+n)} \left(\frac{x}{2}\right)^{2n+\nu}. \tag{8.12a}$$

The infinite series on the right-hand side of (8.12a) is real when x is real; hence, by multiplying this right-hand side by $i^{-\nu}$, we obtain a real function. This suggests the following definition.

Definition 8.4. The modified Bessel function of the first kind of order $\nu \geq 0$ is defined as follows:

$$I_\nu(x) = i^{-\nu} J_\nu(i\,x) = \sum_{n=0}^{\infty} \frac{(-1)^n}{n!\,\Gamma(1+\nu+n)} \left(\frac{x}{2}\right)^{2n+\nu}. \tag{8.13}$$

For $\nu < 0$, the modified Bessel function of the first kind is defined as

$$I_{-\nu}(x) = \sum_{n=0}^{\infty} \frac{1}{n!\,\Gamma(1-\nu+n)} \left(\frac{x}{2}\right)^{2n+\nu}. \tag{8.13a}$$

Remark. If ν is not integer function I_ν and $I_{-\nu}$ are independent, hence the general solution of equation

$$x^2 y'' + x y' - \left(x^2 + \nu^2\right) y = 0, \tag{8.14}$$

can be represented as

$$y = k_1\, I_\nu(x) + k_2\, I_{-\nu}(x).$$

Equation (8.14) is called the Bessel-modified equation of order ν.

Remark. Using modified Bessel functions of the first kind, we can express the general solution of the Airy equation (6.1):

$$y = \sqrt{x}\left(k_1\, I_{\frac{1}{3}}\left(\tfrac{2}{3}\, x^{3/2}\right) + k_2\, I_{-\frac{1}{3}}\left(\tfrac{2}{3}\, x^{3/2}\right)\right). \tag{8.11a}$$

Remark. The comparison between Bessel functions $J_{\pm 1/2}$ and modified functions $I_{\pm 1/2}$ clarifies the analogy we referred to formerly with the trigonometric and hyperbolic functions.

In fact, by repeating, with reasonable adjustments, the technique used in proof of Theorem 7.7, we arrive at

$$I_{1/2}(x) = \sqrt{\frac{2}{\pi x}}\, \sinh x \tag{8.15a}$$

$$I_{-1/2}(x) = \sqrt{\frac{2}{\pi x}}\, \cosh x. \tag{8.15b}$$

8.5 Recurrence relations

Theorem 7.6 can be adapted with no difficulties to provide the recurrence relations for the modified I_ν; it should only be borne in mind that we need to change some sign in (8.16b), (8.16d), and (8.16e):

$$\left(x^\nu\, I_\nu(x)\right)' = x^\nu I_{\nu-1}(x) \tag{8.16a}$$

$$\left(x^{-\nu}\, I_\nu(x)\right)' = x^{-\nu} I_{\nu+1}(x) \tag{8.16b}$$

$$I_\nu'(x) + \frac{\nu}{x}\, I_\nu(x) = I_{\nu-1}(x) \tag{8.16c}$$

$$I_\nu'(x) - \frac{\nu}{x}\, I_\nu(x) = I_{\nu+1}(x) \tag{8.16d}$$

$$I_\nu(x) = \frac{x}{2\nu}\left(I_{\nu-1}(x) - I_{\nu+1}(x)\right) \tag{8.16e}$$

Generating function and integral representation

The considerations that allowed us to define the Bessel functions using the generating function are suitable for the case of modified functions: in fact, it is sufficient to replace x with $i\,x$ and t with $-i\,t$ in (7.27), which leads to

$$g(ix,-it)=\exp\left[\frac{ix}{2}\left(-it+\frac{1}{it}\right)\right]=\exp\left[\frac{x}{2}\left(t+\frac{1}{t}\right)\right].$$

Thus Theorem 7.9 provides

$$\exp\left[\frac{x}{2}\left(t+\frac{1}{t}\right)\right]=\sum_{m=-\infty}^{+\infty}J_m(i\,x)\,(-1)^m\,i^m\,t^m.$$

Finally, recalling definition 8.4 we arrive at the analogous of equation (7.28) for the modified functions:

$$\exp\left[\frac{x}{2}\left(t+\frac{1}{t}\right)\right]=\sum_{m=-\infty}^{+\infty}I_m(x)\,t^m,\ t\neq 0. \tag{8.17}$$

We notice that the analogous of (7.30) is verbatim:

$$I_m(x+y)=\sum_{k=-\infty}^{+\infty}I_k(x)\,I_{m-k}(y). \tag{8.18}$$

To obtain the integral representation, we repeat, with the appropriate adaptations, what was done in Section 7.8. Taking $t=e^{-i\,\vartheta}$ adapting the argument of Theorem 7.10 it is possible to arrive at

$$I_m(x)=\frac{1}{\pi}\int_0^{\pi}e^{x\,\cos\vartheta}\cos(n\,\vartheta)\,\mathrm{d}\vartheta\,. \tag{8.19}$$

8.6 Modified functions of the second kind

Even for the modified functions, when the indices are integers, to obtain the general solution of the equation (8.14), one must take into account the fact that the functions I_m and I_{-m} when $m\in\mathbb{N}$ are not independent; in this situation, they are even coincident, due to relation (7.26). We have

$$I_{-m}=i^m\,J_{-m}(i\,x)=i^m\,(-1)^m\,J_m(i\,x)=i^m\,(-1)^m\,i^m\,I_m(x)=(-1)^{2m}\,I_m(x).$$

To provide a further independent solution of (8.14), it is not customary to use $Y_\nu(i\,x)$; instead, the so-called Macdonald's function (also known as the modified Bessel function of the second kind) is introduced.

Definition. The Macdonald function (modified Bessel function of the second kind) when ν is not an integer is defined as:

$$K_\nu(x) = \frac{\pi}{2\sin(\nu\pi)}\left(I_{-\nu}(x) - I_\nu(x)\right).$$

For $\nu = m$ with m nonnegative integer, the function is defined as:

$$K_m(x) = \lim_{\nu\to m} K_\nu(x).$$

By its construction, K_ν solves (8.14) and is linearly independent of I_ν (this can be verified by direct computation). For the cases of integer order, it is possible to show that

$$K_0(x) = -I_0(x)\left(\ln\frac{x}{2} + \gamma\right) + \sum_{n=1}^{\infty}\frac{(x/2)^{2n}}{(n!)^2}H_n$$

where H_n stands for the n-th harmonic number, i.e. $H_n = 1 + 1/2 + 1/3 + \cdots + 1/n$, while for $m \in \mathbb{N}$

$$K_m(x) = (-1)^{m-1}I_m(x)\ln\frac{x}{2} + \frac{1}{2}\sum_{n=0}^{m-1}(-1)^n\frac{(m-n-1)!}{n!}\left(\frac{x}{2}\right)^{2n-m}$$
$$+\frac{(-1)^m}{2}\sum_{n=0}^{\infty}\frac{(x/2)^{2n+m}}{n!(m+n)!}\left(\psi(m+n+1) + \psi(n+1)\right).$$

Here, γ is the Euler-Mascheroni constant, and ψ stands as usual for the digamma function, see chapter 3.

There are integral representations for the modified K which are proved in [74] or in [113] using nonelementary methods of complex analysis; here, we limit to state them, not before referring the reader to [56] where many integral representations are provided.

$$K_\nu(x) = \frac{1}{2}\left(\frac{x}{2}\right)^\nu\int_0^\infty t^{-(\nu+1)}\exp\left(-t - \frac{x^2}{4t}\right)\mathrm{d}t, \tag{8.20a}$$

$$K_\nu(x) = \int_0^\infty \cosh(\nu t)\,e^{-x\cosh t}\mathrm{d}t. \tag{8.20b}$$

Observe that (8.20b) can also be written, changing variable as $\cosh x = u$ as:

$$K_\nu(x) = \int_1^\infty \frac{e^{-ux}\cosh\left(\nu\,\mathrm{arccosh}\,u\right)}{\sqrt{u^2-1}}\mathrm{d}u. \tag{8.20c}$$

When $\nu = 0$ (8.20c) reduces to

$$K_0(x) = \int_1^\infty \frac{e^{-ux}}{\sqrt{u^2-1}}\mathrm{d}u. \tag{8.20cc}$$

Chapter 9

Gauss hypergeometric function

From a strictly mathematical point of view, the objective of this chapter, through the study of the hypergeometric differential equation, see equation (9.1) further on, is to deal with all second-order differential equations of the type (5.1) where the functions $p(x)$ and $q(x)$ are rational, i.e. quotients of polynomials, and in which three singularities are present. It can be assumed through suitable transformations of variables that these singularities are located in 0, 1 and at infinity. For this reason, we dedicate this chapter to the study of the following equation:

$$x(1-x)\,y'' + [c-(a+b+1)\,x]\,y' - ab\,y = 0, \tag{9.1}$$

where a, b, c are parameters (real or complex) not depending on x, which, as usual, stands for the independent variable of the differential equation.

The fields of application of this equation are, without exaggerating, endless: quantum mechanics, electrodynamics, modern physics, classical mechanics, statistics, economics, and probability, to indicate the main ones. An exhaustive list of applications is nearly impossible, so we limit ourselves to addressing Seaborn's book [104] for the physical sciences, the articles [26, 51, 69, 82] in economics, and Paolella's book [90] for probability. As far as the analytical aspects are concerned, the literature is endless here, too. We highlight [4, 6, 29, 39, 74, 80, 96, 114]. The history of the study of the hypergeometric function is illustrated in [41] and in [105, chapter 1].

9.1 Power series solution around $x = 0$

We look for an analytical solution of the equation (9.1) in a neighbourhood of the origin, a regular singular point of the equation (9.1). Following the Frobenius method, we look for a solution to the form.

$$y = \sum_{n=0}^{\infty} a_n x^{n+s}. \tag{9.2}$$

DOI: 10.1201/9781003171546-9

By omitting the completely standard details, we arrive at

$$a_0 s(s+c-1) + \sum_{n=0}^{\infty} a_{n+1}(n+1+s)(n+s+c)x^{n+s}$$
$$-\sum_{n=0}^{\infty} a_n \left[(n+s)(n+s+a+b)+a\,b\right] x^{n+s} = 0.$$

Consequently, we obtain the indicial equation:

$$s(s+c-1)a_0 = 0 \tag{9.3}$$

and difference equation satisfied by the sequence (a_n)

$$a_{n+1} = \frac{(n+s)(n+s+a+b)+ab}{(n+1+s)(n+s+c)}\, a_n. \tag{9.4}$$

Now, in (9.3), we cannot assume $a_0 = 0$ since, due to (9.4), we would have $a_n = 0$ for any $n \in \mathbb{N}$, which leads to the trivial solution. The solutions to (9.3) are $s = 0$ and $s = 1-c$. We start with $s = 0$, inserting in (9.4) we obtain:

$$\begin{aligned}
a_1 &= a_0 \frac{ab}{c} \\
a_2 &= a_0 \frac{a(a+1)b(b+1)}{1\cdot 2\cdot c(c+1)} \\
a_3 &= a_0 \frac{a(a+1)(a+2)b(b+1)(b+2)}{1\cdot 2\cdot 3\cdot c(c+1)(c+2)}. \\
&\cdots\cdots
\end{aligned}$$

It is more expressive to use Pochhamer symbols $(a)_n$, $(b)_n$, $(c)_n$ where $n \in \mathbb{N}$, so that, choosing $a_0 = 1$, meaning that we impose the initial condition $y(0) = 1$, (9.2) becomes:

$${}_2\mathrm{F}_1\left(\left.\begin{matrix} a, b \\ c \end{matrix}\right| x\right) := \sum_{n=0}^{\infty} \frac{(a)_n\,(b)_n}{(c)_n}\,\frac{x^n}{n!}. \tag{9.5}$$

The notation used on the left-hand side of (9.5) emphasises that the solution of (9.1) is influenced by parameters a, b, $c \in \mathbb{C}$. Moreover, this particular grouping is because a and b are interchangeable without altering the solution due to the form of (9.1) itself. Notice that in this manner, we have detected the integral of (9.1) satisfying the initial conditions $y(0) = 1$ as well $y'(0) = (ab)/c$.

To determine the general solution of the equation (9.1), we need the second solution of the equation (9.3). Such a solution is generated by the choice $s = 1-c$. Therefore, the second independent solution of (9.1) is found as:

$$x^{1-c}\,{}_2\mathrm{F}_1\left(\left.\begin{matrix} 1+a-c, 1+b-c \\ 2-c \end{matrix}\right| x\right).$$

Recalling that, we must assume that c is not a positive integer greater or equal to 2. For the complete treatment of the equation, we refer to [105], where the remaining values of c and the solution at the other singular points $x = 1$ and $x = \infty$ are examined, obtaining the famous 24 Kummer solutions.

Remark. It is worth noting that some authors use other notations. Our notation is due to Appell [11], while Bailey [18] writes $F(a, b; c; x)$.

Power series solution around $x = 1$ and $x = \infty$

This section shows examples of independent variable transformation that mutates the hypergeometric equation (9.1) into an equation of the same type but with different coefficients. These transformations are relevant in classifying all possible solutions of (9.1), presented in [64, 97, 105]. The principal contribution to the individuation of all solutions is due to Kummer, who identified twenty-four solutions of (9.1), using fractional linear transformations of the independent variable of the form:

$$u = 1 - x, \quad u = \frac{1}{x}, \quad u = \frac{x}{x-1}, \quad u = \frac{1}{1-x}.$$

In this process, the transformation rules (5.18) are extremely useful. Moreover, notice that these transformations map the singularity set $\{0, 1, \infty\}$ of (9.1) in itself and that the new differential equation obtained in this process is of the same type of (9.1). The approach transforms the independent variable x so that the new variable u lies in the set $|u| < 1$ and then uses the solution procedures outlined in Section 9.1.

We will omit the details, addressing to [92] or [105] Section 1.3 for the extensive treatment; here, we content ourselves with pointing out that at the end of these transformations, we find, in the new variables, differential equations of a hypergeometric type whose solutions are represented by ${}_2\mathrm{F}_1$ in which the parameters and the independent variable are appropriately changed.

9.2 Properties of the hypergeometric series

Function ${}_2\mathrm{F}_1$ is called "hypergeometric function". Many authors add the name of Gauss when referring to ${}_2\mathrm{F}_1$, as we will do, even if, as a matter of fact, the process that led to the study of the differential equation (9.1) started much before and has many eminent contributors, such as Wallis, Stirling, Euler, and Pfaff: the historical reconstruction about the investigations regarding ${}_2\mathrm{F}_1$ is magistrally provided in [41]. But Gauss indeed was the first to use the term "hypergeometric" addressing the fact that when in (9.1) we assume $a = b =$

$c = 1$, then the power series ${}_2\mathrm{F}_1$ is indeed the geometric series, as:

$$ {}_2\mathrm{F}_1\left(\left.\begin{matrix}1,1\\1\end{matrix}\right| x\right) = \sum_{n=0}^{\infty} x^n. $$

Hypergeometric was introduced to remind us that this procedure recognises the geometric series as a particular case.

If $a = -m$ or $b = -m$ with $m \in \mathbb{N}$ the series terminates and ${}_2\mathrm{F}_1$ reduces to a (hypergeometric) polynomial. This is due to the properties of Pochhammer symbols, since if $m < n$, then $(-m)_n = 0$, see proposition 2.7. Moreover, using formulæ (2.19) and (2.19a), we can state the following:

Theorem 9.1. If $m \in \mathbb{N}$ and, b, $c \in \mathbb{C}$, then

$$ \begin{aligned} {}_2\mathrm{F}_1\left(\left.\begin{matrix}-m,b\\c\end{matrix}\right| x\right) &= \sum_{n=0}^{m} (-1)^n \frac{m!}{(m-n)!} \frac{(b)_n}{(c)_n} \frac{x^n}{n!}, \\ &= \sum_{n=0}^{m} (-1)^n \binom{m}{n} \frac{(b)_n}{(c)_n} x^n. \end{aligned} $$

Theorem. The radius of convergence of series (9.5) is one unless a or b are negative integers. At the boundary of the convergence circle $|x| = 1$, series (9.5) converges for $\mathrm{Re}\,(a+b) < \mathrm{Re}\,c$ and diverges for $\mathrm{Re}\,(a+b) \geq \mathrm{Re}\,c$.

Proof. Instead, if neither a nor b are negative integers, the radius of convergence is obtained by evaluating:

$$ \lim_{n\to\infty} \frac{(a)_n(b)_n}{(c)_n} \frac{(c)_{n+1}}{(a)_{n+1}(b)_{n+1}} = \lim_{n\to\infty} \frac{(c+n)(n+1)}{(a+n)(b+n)} = 1. $$

At the boundary of the convergence circle $|x| = 1$, we refer to [96, page 46] for the general treatment; here, we limit to study the behaviour of the series, assuming a, b, $c > 0$ using the Raabe criterion, which means that, denoting with x_n the term of the series (9.1) evaluated for $x = 1$, we have to consider the following limit for $n \to \infty$

$$ q = \lim_{n\to\infty} n\left(\frac{x_n}{x_{n+1}} - 1\right) = \lim_{n\to\infty} n\left(\frac{(n+1)(c+n)}{(a+n)(b+n)} - 1\right) = 1 - a - b + c, $$

and recalling that for $q > 1$, the series is convergent, the thesis follows. □

Proposition 9.2. For $|x| < 1$ the derivative of ${}_2\mathrm{F}_1$ is given by

$$ \frac{\mathrm{d}}{\mathrm{d}x}\, {}_2\mathrm{F}_1\left(\left.\begin{matrix}a,b\\c\end{matrix}\right| x\right) = \frac{ab}{c}\, {}_2\mathrm{F}_1\left(\left.\begin{matrix}1+a,1+b\\1+c\end{matrix}\right| x\right). \tag{9.7} $$

Proof. Differentiating term by term (9.5) we find out

$$ \frac{\mathrm{d}}{\mathrm{d}x}\, {}_2\mathrm{F}_1\left(\left.\begin{matrix}a,b\\c\end{matrix}\right| x\right) = \sum_{n=1}^{\infty} \frac{(a)_n(b)_n}{(c)_n} \frac{x^{n-1}}{(n-1)!}. $$

Or, after indexes rearrangement:

$$\frac{\mathrm{d}}{\mathrm{d}x}\,{}_2\mathrm{F}_1\left(\begin{matrix}a,b\\c\end{matrix}\middle|\,x\right)=\sum_{n=0}^{\infty}\frac{(a)_{n+1}(b)_{n+1}}{(c)_{n+1}}\,\frac{x^n}{n!}\,.$$

Formula (9.7) now follows from proposition 2.5, formula (2.16) (see page 63).

□

Remark. Many usual functions, in addition to the geometric series, can be expressed in terms of the hypergeometric series; for example:

$$\frac{\arcsin x}{x}={}_2\mathrm{F}_1\left(\begin{matrix}\frac{1}{2},\frac{1}{2}\\\frac{3}{2}\end{matrix}\middle|\,x^2\right),\qquad \frac{\arctan x}{x}={}_2\mathrm{F}_1\left(\begin{matrix}\frac{1}{2},1\\\frac{3}{2}\end{matrix}\middle|\,-x^2\right),$$

$$\frac{\ln(1-x)}{x}=-\,{}_2\mathrm{F}_1\left(\begin{matrix}1,1\\2\end{matrix}\middle|\,x\right),\qquad \frac{\ln(1+x)}{x}={}_2\mathrm{F}_1\left(\begin{matrix}1,1\\2\end{matrix}\middle|\,-x\right).$$

9.3 Integral representations

There are several ways to represent ${}_2\mathrm{F}_1$ using integrals. This section presents the most popular representation theorem, which we will address as "integral representation theorem". We briefly summarise [41, page 26] for the historical reconstruction. Euler came close to the integral representation we are about to present, and Legendre was the first to provide an integral representation for the hypergeometric function. Still, the following formula (9.8) appeared first in the doctoral dissertation of Vorsselman de Heer in 1833.

Theorem 9.3. If $\operatorname{Re} c > \operatorname{Re} a > 0$ and if $|x| < 1$ then

$${}_2\mathrm{F}_1\left(\begin{matrix}a,b\\c\end{matrix}\middle|\,x\right)=\frac{\Gamma(c)}{\Gamma(a)\Gamma(c-a)}\int_0^1\frac{u^{a-1}(1-u)^{c-a-1}}{(1-xu)^b}\,\mathrm{d}u \tag{9.8}$$

Proof. We use the binomial series expansion, see (2.22) at page 65:

$$(1-u)^{-\alpha}=\sum_{n=0}^{\infty}\frac{(\alpha)_n}{n!}u^n\,.$$

Given this, we can evaluate the integral with the right side of (9.8) by integrating by series:

$$\int_0^1\frac{u^{a-1}\,(1-u)^{c-a-1}}{(1-u\,x)^b}\,\mathrm{d}u=\sum_{m=0}^{\infty}\int_0^1 u^{a-1}\,(1-u)^{c-a-1}\,\frac{(b)_m}{m!}u^m x^m\mathrm{d}u$$

$$=\sum_{m=0}^{\infty}\frac{(b)_m}{m!}x^m\int_0^1 u^{a+m-1}\,(1-u)^{c-a-1}\,\mathrm{d}u$$

$$=\sum_{m=0}^{\infty}\frac{(b)_m}{m!}x^m\,\frac{\Gamma(a+m)\Gamma(c-a)}{\Gamma((a+m)+(c-a))}.$$

Recalling that $\Gamma(a+m) = (a)_m\Gamma(a)$, $\Gamma(c+m) = (c)_m\Gamma(c)$ we eventually get (9.8). □

The integral representation (9.8), as well as its dual (9.8a) presented below, allows through appropriate changes of the variable to obtain further representations of ${}_2\mathrm{F}_1$. Here, we limit ourselves to presenting those deriving from the representation obtained in the Theorem 9.3.

Corollary. If $\operatorname{Re} c > \operatorname{Re} a > 0$ and if $|x| < 1$ then

$$ {}_2\mathrm{F}_1\left(\begin{matrix} a,b \\ c \end{matrix}\middle| x\right) = \frac{\Gamma(c)}{\Gamma(a)\Gamma(c-a)} \int_0^\infty \frac{s^{a-1}(s+1)^{b-c}}{(1+s(1-x))^b}\, ds \tag{9.9a} $$

$$ {}_2\mathrm{F}_1\left(\begin{matrix} a,b \\ c \end{matrix}\middle| x\right) = \frac{\Gamma(c)}{\Gamma(a)\Gamma(c-a)} \int_1^\infty \frac{(s-1)^{c-a-1}s^{b-c}}{(s-x)^b}\, ds \tag{9.9b} $$

$$ {}_2\mathrm{F}_1\left(\begin{matrix} a,b \\ c \end{matrix}\middle| x\right) = \frac{\Gamma(c)}{\Gamma(a)\Gamma(c-a)} \int_0^\infty \frac{e^{-as}\left(1-e^{-s}\right)^{c-a-1}}{\left(1-e^{-s}x\right)^b}\, ds \tag{9.9c} $$

Proof. All formulas follow from (9.8): formula (9.9a) putting $u = \frac{s}{1+s}$, while for (9.9b) we use $u = \frac{1}{s}$ and, finally, $u = e^{-s}$ produces (9.9c). □

Remark. The integral representation formula (9.8) allows the hypergeometric function to be extended, of course only for $\operatorname{Re} c > \operatorname{Re} a > 0$, to the whole complex plane, without the subset of the real axis consisting of the half-line made up of the points $(x, 0)$ with $x \geq 1$. A detailed rigorous justification of this statement is provided by [74, pages 239–240].

Remark. Recalling that parameters a and b in the power series (9.5) interchanges, this property also holds for (9.8), hence we infer a second integral representation formula, which holds for $\operatorname{Re} c > \operatorname{Re} b > 0$ and $|x| < 1$:

$$ {}_2\mathrm{F}_1\left(\begin{matrix} a;b \\ c \end{matrix}\middle| x\right) = \frac{\Gamma(c)}{\Gamma(c-b)\Gamma(b)} \int_0^1 \frac{u^{b-1}(1-u)^{c-b-1}}{(1-x\,u)^a}\, du\,. \tag{9.8a} $$

The equality of the two integrals in (9.8) and in (9.8a), when both conditions $\operatorname{Re} c > \operatorname{Re} a > 0$ and $\operatorname{Re} c > \operatorname{Re} b > 0$ hold, is not obvious to verify directly, it can be proved using the double integral of Erdély, see [47].

$$ \frac{\Gamma^2(c)}{\Gamma(a)\Gamma(b)\Gamma(c-a)\Gamma(c-b)} \int_0^1\int_0^1 \frac{u^{b-1}v^{a-1}(1-u)^{c-b-1}(1-v)^{c-a-1}}{(1-uvt)^c} du\, dv. $$

9.4 Application of Theorem 9.3

The integral representation formulæ (9.8), and (9.8a) allow the calculation of integrals that cannot be expressed using elementary functions. When we use the verb, calculate, we can represent a given integral through a particular occurrence of the hypergeometric series. We will limit ourselves to presenting a few examples, as preparing a complete table of integrals of a certain type is outside our scope.

Elliptic integrals

This typology of integrals is known as elliptic integrals, which were studied by Legendre, Jacobi and Weierstrass, and is treated by introducing a class of functions called elliptic functions, [28, 25], which we will not present here. These are integrals of the form

$$\int R\left(x, \sqrt{p(x)}\right) \, \mathrm{d}x \, .$$

Here, R is a rational function and $p(x)$ a polynomial of degree three or four. Here, we are going to give a brief mention of their hypergeometric treatment in some particular cases.

Given three real numbers $\alpha < \beta < \gamma$, consider the integral

$$I(\alpha, \beta, \gamma) = \int_{\alpha}^{\beta} \frac{\mathrm{d}x}{\sqrt{(x-\alpha)(\beta-x)(\gamma-x)}} .$$

The computational strategy is extremely naive and simple: it is a matter of normalising, with the change of variable $x = \alpha + (\beta - \alpha)\, t$, the assigned interval of integration $[\alpha, \beta]$, to obtain the interval $[0, 1]$. Thus, we obtain the following:

$$I(\alpha, \beta, \gamma) = \int_{0}^{1} \frac{\mathrm{d}t}{\sqrt{t(1-t)(\gamma-\alpha-(\beta-\alpha)\, t)}} .$$

Finally, to recognise the structure of (9.8) we only need to factor $\gamma - \alpha$:

$$I(\alpha, \beta, \gamma) = \frac{1}{\sqrt{\gamma-\alpha}} \int_{0}^{1} \frac{t^{-1/2}(1-t)^{-1/2}}{\left(1 - \frac{\beta-\alpha}{\gamma-\alpha}\, t\right)^{1/2}} \, \mathrm{d}t.$$

Hence, always referring to (9.8), the hypergeometric parameters of the integral are finally detected as:

$$a - 1 = -\frac{1}{2}, \, c - a - 1 = -\frac{1}{2}, \, b = \frac{1}{2} \implies a = \frac{1}{2}, \, c = 1 \implies \frac{\Gamma(c)}{\Gamma(a)\Gamma(c-a)} = \frac{1}{\pi} .$$

Thus, we finally infer

$$I(\alpha, \beta, \gamma) = \frac{\pi}{\sqrt{\gamma-\alpha}} \, {}_2\mathrm{F}_1\left(\begin{matrix} 1/2, 1/2 \\ 1 \end{matrix} \middle| \, \frac{\beta-\alpha}{\gamma-\alpha} \right) .$$

Using the same technique, always in the hypothesis in which $a < b < c$, the following relations are found:

$$\int_a^b \sqrt{\frac{c-x}{(b-x)(x-a)}}\,\mathrm{d}x = \pi\sqrt{c-a}\,{}_2\mathrm{F}_1\left(\begin{matrix}\frac{1}{2}, -\frac{1}{2}\\ 1\end{matrix}\,\middle|\,\frac{b-a}{c-a}\right),$$
$$\int_a^b \sqrt{\frac{b-x}{(c-x)(x-a)}}\,\mathrm{d}x = \frac{\pi}{2}\frac{b-a}{\sqrt{c-a}}\,{}_2\mathrm{F}_1\left(\begin{matrix}\frac{1}{2}, \frac{1}{2}\\ 2\end{matrix}\,\middle|\,\frac{b-a}{c-a}\right),$$
$$\int_a^b \sqrt{\frac{x-a}{(c-x)(b-x)}}\,\mathrm{d}x = \frac{\pi}{2}\frac{b-a}{\sqrt{c-a}}\,{}_2\mathrm{F}_1\left(\begin{matrix}\frac{1}{2}, \frac{3}{2}\\ 2\end{matrix}\,\middle|\,\frac{b-a}{c-a}\right).$$

A generalised Eulerian integral

In the following particular case, Theorem 9.3 allows to evaluate integrals very similar to Eulerian integrals of the first kind.

Theorem. If $a > 1$ and $b \in \mathbb{R}$ then

$$\int_0^1 x^{a-1}(1+x)^b\mathrm{d}x = \frac{1}{a}\,{}_2\mathrm{F}_1\left(\begin{matrix}a; -b\\ 1+a\end{matrix}\,\middle|\,-1\right).$$

Proof. Just take in the formula (9.8) of Theorem 9.3 $a = a$, $b = -b$, $c = a+1$. □

Incomplete beta function

The incomplete beta function is defined as:

$$\mathrm{B}(x;a,b) = \int_0^x t^{a-1}(1-t)^{b-1}\,\mathrm{d}t, \tag{9.10}$$

where $\mathrm{Re}\,a > 0$, $\mathrm{Re}\,b > 0$, $0 \le x \le 1$. The incomplete beta can be represented in hypergeometric terms; in fact, we can prove the following theorem.

Theorem.

$$\mathrm{B}(x;a,b) = \frac{x^a}{a}\,{}_2\mathrm{F}_1\left(\begin{matrix}a, 1-b\\ a+1\end{matrix}\,\middle|\,x\right). \tag{9.11}$$

Proof. Changing variables in (9.10) as $t = xu$ we obtain

$$\mathrm{B}(x;a,b) = \int_0^1 x^{a-1}u^{a-1}(1-xu)^{b-1}x\mathrm{d}u = x^a\int_0^1 \frac{u^{a-1}}{(1-xu)^{1-b}}\mathrm{d}u. \tag{9.12}$$

On the right-hand side of (9.12), we can invoke Theorem 9.3, proving (9.11). □

9.5 Behaviour at the boundary of the convergence disk

Because the hypergeometric series converges for $|x| < 1$, it is of great interest, not only for its theoretical implications but also for its applications, to examine its behaviour at the boundary of the convergence disk. Even with a certain loss of generality, because its use requires assumptions about hypergeometric parameters, we will use the integral representation theorem, which allows more elementary proofs. We refer to the above [4, 6, 18, 96, 104, 105] for a more general treatment. We begin with a result due to S. Ramanujan.

Theorem. For any $x \in]0, 1[$ we have

$$ {}_2\mathrm{F}_1\left(\begin{array}{c} \frac{1}{2}; x \\ \frac{3}{2} \end{array}\middle| 1\right) = \frac{\sqrt{\pi}}{2}\frac{\Gamma(1-x)}{\Gamma\left(\frac{3}{2}-x\right)}. \tag{9.13} $$

Proof. We use the integral representation formula (9.8) taking $a = 1/2$, $b = x$, $c = 3/2$ and $x = 1$, since the relevant integral converges.

$$ \begin{aligned} {}_2\mathrm{F}_1\left(\begin{array}{c} \frac{1}{2}, x \\ \frac{3}{2} \end{array}\middle| 1\right) &= \frac{\Gamma\left(\frac{3}{2}\right)}{\Gamma(x)\Gamma\left(\frac{3}{2}-x\right)}\int_0^1 \frac{s^{x-1}(1-s)^{\frac{3}{2}-x-1}}{(1-s)^{\frac{1}{2}}}\mathrm{d}s \\ &= \frac{\sqrt{\pi}}{2}\frac{1}{\Gamma(x)\Gamma\left(\frac{3}{2}-x\right)}\int_0^1 s^{x-1}(1-s)^{(1-x)-1}\mathrm{d}s \\ &= \frac{\sqrt{\pi}}{2}\frac{1}{\Gamma(x)\Gamma\left(\frac{3}{2}-x\right)}\mathrm{B}\left(x, 1-x\right) \\ &= \frac{\sqrt{\pi}}{2}\frac{1}{\Gamma(x)\Gamma\left(\frac{3}{2}-x\right)}\frac{\Gamma(x)\Gamma(1-x)}{\Gamma(1)} \end{aligned} $$

Simplifying, we find (9.13). □

The following result is known as Gauss' summation formula, in which the value of the hypergeometric function is calculated at the singular point $x = 1$ under suitable assumptions on the parameters a, b, c.

Theorem 9.4. If $c > a + b$ then

$$ {}_2\mathrm{F}_1\left(\begin{array}{c} a; b \\ c \end{array}\middle| 1\right) = \frac{\Gamma(c)\Gamma(c-a-b)}{\Gamma(c-a)\Gamma(c-b)}. \tag{9.14} $$

Proof. For simplicity, we give the proof by making the additional hypothesis $c > b > 0$, referring, for example, to [39, pages 352–353] for the general case. The additional hypothesis allows us to use the integral representation (9.8):

$$ \lim_{x\to 1^-} {}_2\mathrm{F}_1\left(\begin{array}{c} a; b \\ c \end{array}\middle| x\right) = \lim_{x\to 1^-} \frac{\Gamma(c)}{\Gamma(b)\Gamma(c-b)}\int_0^1 \frac{s^{b-1}(1-s)^{c-b-1}}{(1-x\,s)^a}\,\mathrm{d}s, $$

which, since $c > a + b$ ensures the convergence of the hypergeometric series:

$$
\begin{aligned}
{}_2\mathrm{F}_1\left(\begin{matrix} a; b \\ c \end{matrix} \middle| 1\right) &= \frac{\Gamma(c)}{\Gamma(b)\Gamma(c-b)} \int_0^1 s^{b-1}(1-s)^{c-a-b-1}\,\mathrm{d}s \\
&= \frac{\Gamma(c)}{\Gamma(b)\Gamma(c-b)}\mathrm{B}(b, c-a-b) = \frac{\Gamma(c)}{\Gamma(b)\Gamma(c-b)} \frac{\Gamma(b)\Gamma(c-a-b)}{\Gamma(b+c-a-b)} \\
&= \frac{\Gamma(c)\Gamma(c-a-b)}{\Gamma(c-a)\Gamma(c-b)}.
\end{aligned}
$$

□

Gauss' summation formula allows us to prove an important result, the Chu Vandermonde identity.

Theorem. Given $m \in \mathbb{N}$ if $c > a - m$ then

$$
{}_2\mathrm{F}_1\left(\begin{matrix} -m; a \\ c \end{matrix} \middle| 1\right) = \frac{(c-a)_m}{(c)_m}. \tag{9.15}
$$

Proof. Identity (9.15) follows from (9.14), in fact since we have

$$
{}_2\mathrm{F}_1\left(\begin{matrix} -m; a \\ c \end{matrix} \middle| 1\right) = \frac{\Gamma(c)\Gamma(c+m-a)}{\Gamma(c+m)\Gamma(c-a)}.
$$

Recalling the definition of Pochhamer symbol we have that $\Gamma(c+m-a) = (c-a)_m\Gamma(c-a)$ and $\Gamma(c+m) = (c)_m\Gamma(c)$, thus inserting in the last equality, thesis follows. □

Another consequence of the integral representation formula (9.8) is the Kummer summation formula. Here, we give the proof proposed in [83], even if this proof works in a less general framework compared with the proof provided in [96].

Theorem 9.5. If $1 + a - b \neq 0$ is not a negative integer and if $\operatorname{Re} b < 1$, then

$$
{}_2\mathrm{F}_1\left(\begin{matrix} a; b \\ 1+a-b \end{matrix} \middle| -1\right) = \frac{\Gamma(1+a-b)\,\Gamma(1+\frac{a}{2})}{\Gamma(1+a)\,\Gamma(1+\frac{a}{2}-b)} \tag{9.16}
$$

Proof. Our simplified proof also assumes $a > 0$, and it is based on the convergent integral:

$$
A_2(a,b) = \int_0^1 \frac{t^{a-1}}{(1-t^2)^b}\,\mathrm{d}t. \tag{9.17}
$$

Changing variable in (9.17) as $t^2 = u$; we find

$$
A_2(a,b) = \frac{1}{2}\frac{\Gamma\left(\frac{a}{2}\right)\Gamma(1-b)}{\Gamma\left(1+\frac{a}{2}-b\right)}.
$$

On the other side, writing (9.17) as

$$A_2(a,b) = \int_0^1 \frac{t^{a-1}(1-t)^{-b}}{(1+t)^b}\mathrm{d}t; \tag{9.18}$$

from (9.8) we infer:

$$A_2(a,b) = \frac{\Gamma(a)\Gamma(1-b)}{\Gamma(1-a-b)}\,{}_2\mathrm{F}_1\left(\begin{array}{c} a;b \\ 1+a-b \end{array}\middle| -1\right). \tag{9.19}$$

Comparing the two expressions found for $A_2(a,b)$ and recalling that

$$\frac{\Gamma\left(\frac{a}{2}\right)}{2\Gamma(a)} = \frac{\Gamma\left(1+\frac{a}{2}\right)}{\Gamma(1+a)}$$

we arrive at (9.16).

□

9.6 Some evaluations inside the convergence disk

Let's start with a result from Bailey; see [18].

Theorem. If $a < 1$, $a + c > 1$ then

$${}_2\mathrm{F}_1\left(\begin{array}{c} a;1-a \\ c \end{array}\middle| \frac{1}{2}\right) = \frac{2^{1-c}\sqrt{\pi}\,\Gamma(c)}{\Gamma(\frac{a+c}{2})\,\Gamma(\frac{c-a+1}{2})}. \tag{9.20}$$

Proof. If in (9.8) we take $b = 1-a$ and $x = 1/2$ we see that:

$${}_2\mathrm{F}_1\left(\begin{array}{c} a;1-a \\ c \end{array}\middle| \frac{1}{2}\right) = \frac{\Gamma(c)}{\Gamma(1-a)\Gamma(c+a-1)}\int_0^1 \frac{(1-s)^{c+a-2}}{(2s-s^2)^a}\mathrm{d}s.$$

Changing the variable in the integral on the right-hand side as $(1-s)^2 = u$, we obtain:

$$\begin{aligned} {}_2\mathrm{F}_1\left(\begin{array}{c} a;1-a \\ c \end{array}\middle| \frac{1}{2}\right) &= \frac{2^{a-1}\Gamma(c)}{\Gamma(1-a)\Gamma(c+a-1)}\int_0^1 (1-u)^{-a}u^{\frac{1}{2}(a+c-3)}\mathrm{d}u \\ &= \frac{2^{a-1}\Gamma(c)}{\Gamma(1-a)\Gamma(c+a-1)}\frac{\Gamma(1-a)\Gamma\left(\frac{1}{2}(a+c-1)\right)}{\Gamma\left(\frac{1}{2}(-a+c+1)\right)}. \end{aligned}$$

Thesis the follows from the duplication formula (2.50) taking $x = \frac{a+c-1}{2}$.

□

Remark. The formula (9.20) is often reported in the more elegant form:

$$ {}_2\mathrm{F}_1\left(\begin{matrix} a; 1-a \\ c \end{matrix} \middle| \frac{1}{2} \right) = \frac{\Gamma(\frac{c}{2})\,\Gamma(\frac{c+1}{2})}{\Gamma(\frac{a+c}{2})\,\Gamma(\frac{c-a+1}{2})}. \tag{9.20b} $$

which is obtained by further using the duplication formula (2.50) in (9.20) having observed that

$$ \Gamma(c) = \frac{2^{c-1}}{\sqrt{\pi}}\,\Gamma(\tfrac{c}{2})\,\Gamma(\tfrac{c+1}{2}). $$

We now focus on an interesting evaluation of ${}_2\mathrm{F}_1$ still for $x = 1/2$ and for $a = b = 1$. Considering a particular situation results in a surprising connection with the digamma function. Moreover, this evaluation will be useful in the computation of a hard definite integral, see formula (11.29) we will present in chapter 11.

Theorem. If $c > 1$ then

$$ {}_2\mathrm{F}_1\left(\begin{matrix} 1; 1 \\ c \end{matrix} \middle| \frac{1}{2} \right) = (c-1)\left(\psi\left(\frac{c}{2}\right) - \psi\left(\frac{c-1}{2}\right)\right). \tag{9.21} $$

Proof. We express the left-hand side of (9.21) using the integral representation Theorem 9.3, in this case the integral (9.8) yields:

$$ {}_2\mathrm{F}_1\left(\begin{matrix} 1; 1 \\ c \end{matrix} \middle| \frac{1}{2} \right) = (c-1)\int_0^1 \frac{(1-u)^{c-2}}{1-\frac{1}{2}u}\,\mathrm{d}u = 2\,(c-1)\int_0^1 \frac{v^{c-2}}{1+v}\,\mathrm{d}v. \tag{$\bullet$} $$

Notice that to arrive at ($\bullet$) we change the variable as $u = 1 - v$.

On the other side, recalling proposition 3.2 equation (3.9) we have, after some computations:

$$ \begin{aligned} \psi\left(\frac{c}{2}\right) - \psi\left(\frac{c-1}{2}\right) &= \int_0^1 \frac{u^{\frac{c-1}{2}-1} - u^{\frac{c}{2}-1}}{1-u}\,\mathrm{d}u \\ &= \int_0^1 \frac{u^{\frac{c}{2}-1}\left(u^{-\frac{1}{2}} - 1\right)}{1-u}\,\mathrm{d}u \\ &= \int_0^1 \frac{u^{\frac{c}{2}-\frac{3}{2}}\left(1 - u^{\frac{1}{2}}\right)}{1-u}\,\mathrm{d}u. \end{aligned} $$

Now changing the variable in the last integral as $u^{\frac{1}{2}} = v$ yields:

$$ \psi\left(\frac{c}{2}\right) - \psi\left(\frac{c-1}{2}\right) = 2\int_0^2 \frac{v^{c-2}}{v+1}\,\mathrm{d}v. \tag{$\bullet\bullet$} $$

Identity (9.21) follows by comparing ($\bullet$) and ($\bullet\bullet$). □

The following couple of evaluations of the hypergeometric series are due to M. Spiegel, see [106].

Theorem. If $p > 0$ then:

$$ {}_2\mathrm{F}_1\left(\begin{array}{c|c} \frac{1}{2}; p & 3 \\ 3p & \overline{4} \end{array} \right) = \left(\frac{16}{27}\right)^p \frac{\Gamma(3p)\Gamma(p)}{\Gamma^2(2p)}. \tag{9.22a} $$

$$ {}_2\mathrm{F}_1\left(\begin{array}{c|c} 1-p; p & 1 \\ 3p & \overline{2} \end{array} \right) = \left(\frac{1}{2}\right)^p \frac{\Gamma(3p)\Gamma(p)}{\Gamma^2(2p)}. \tag{9.22b} $$

Proof. Choose $p > 0$ and observe that, due to the symmetry of the graph of the function $\vartheta \mapsto \sin^{2p-1}(3\vartheta)$, we have that:

$$ \int_0^{\pi} \sin^{2p-1}(3\vartheta)\,\mathrm{d}\vartheta = \int_0^{\frac{\pi}{3}} \sin^{2p-1}(3\vartheta)\,\mathrm{d}\vartheta. \tag{9.23} $$

Hence, performing the substitution $\varphi = 3\vartheta$ we obtain

$$ \int_0^{\pi} \sin^{2p-1}(3\vartheta)\,\mathrm{d}\vartheta = \frac{1}{3}\int_0^{\pi} \sin^{2p-1}\varphi\,\mathrm{d}\varphi = \frac{2}{3}\int_0^{\frac{\pi}{2}} \sin^{2p-1}\varphi\,\mathrm{d}\varphi. \tag{9.24} $$

Notice that the last inequality is again due to the symmetry property of the sine function. Now observe that the integral on the right-hand side of (9.24) can be evaluated in terms of beta. Then the gamma function, see page 67, equation (2.25a), arriving, also recalling (9.23), at:

$$ \int_0^{\frac{\pi}{3}} \sin^{2p-1}(3\vartheta)\,\mathrm{d}\vartheta = \frac{\sqrt{\pi}}{3}\frac{\Gamma(p)}{\Gamma(p+\frac{1}{2})}. \tag{9.24a} $$

Now in the integral on the left-hand side of (9.24a), we use the triplication formula for the sine function, i.e. $\sin(3\varphi) = 3\sin\varphi - 4\sin^3\varphi$; hence, we can express (9.24a) as:

$$ \int_0^{\frac{\pi}{3}} \sin^{2p-1}\vartheta\left(3 - 4\sin^2\varphi\right)\mathrm{d}\vartheta = \frac{\sqrt{\pi}}{3}\frac{\Gamma(p)}{\Gamma(p+\frac{1}{2})}. \tag{9.24b} $$

We next change variables to the left integral in (9.24b), placing $4\sin^2\varphi = 3u$ so that the new integration limits are $u = 0$, corresponding to $\varphi = 0$ and $u = 1$ corresponding to $\varphi = \pi/3$. Having observed that $\vartheta = \arcsin\left(\sqrt{\frac{3}{4}u}\right)$ the hypergeometric nature of this integral is revealed, and we arrive at

$$ \frac{3^{3p-1}}{2^{2p+1}}\int_0^1 \frac{u^{p-1}(1-u)^{2p-1}}{\left(1-\frac{3}{4}u\right)^{1/2}}\,\mathrm{d}\vartheta = \frac{\sqrt{\pi}}{3}\frac{\Gamma(p)}{\Gamma(p+\frac{1}{2})}. \tag{9.24c} $$

We can now use (9.8), in (9.24c) obtaining:

$$ \frac{3^{3p-1}}{2^{2p+1}}\frac{\Gamma(p)\Gamma(2p)}{\Gamma(3p)}\,{}_2\mathrm{F}_1\left(\begin{array}{c|c} \frac{1}{2}; p & 3 \\ 3p & \overline{4} \end{array} \right) = \frac{\sqrt{\pi}}{3}\frac{\Gamma(p)}{\Gamma(p+\frac{1}{2})}. \tag{9.24d} $$

Solving for the hypergeometric in (9.24d) we get

$$ {}_2F_1\left(\begin{matrix}\frac{1}{2};p\\ 3p\end{matrix}\,\middle|\,\frac{3}{4}\right)=\frac{2^{2p+1}}{3^{3p}}\sqrt{\pi}\,\frac{\Gamma(3p)}{\Gamma(2p)\,\Gamma(p+\frac{1}{2})}. \tag{9.24e} $$

Finally expressing $\Gamma(p+\frac{1}{2})$ with the duplication formula (2.50), written as:

$$ \Gamma(p+\tfrac{1}{2})=2^{1-2p}\,\pi^{\frac{1}{2}}\,\frac{\Gamma(2p)}{\Gamma(p)}. $$

Formula (9.22a) follows.

The strategy for proving (9.22b) is quite similar and is based on the quadruplication of the sine and the integral:

$$ I_4=\int_0^{\frac{\pi}{4}}\sin^{2p-1}(4\vartheta)\,\mathrm{d}\vartheta. $$

Thus, we have

$$ I_4=\frac{1}{4}\int_0^{\pi}\sin^{2p-1}\vartheta\,\mathrm{d}\vartheta=\frac{1}{2}\int_0^{\frac{\pi}{2}}\sin^{2p-1}\vartheta\,\mathrm{d}\vartheta=\frac{\sqrt{\pi}}{4}\,\frac{\Gamma(p)}{\Gamma(p+\frac{1}{2})}. \tag{9.25} $$

On the other side, using the identity $\sin(4\vartheta)=4\sin\vartheta\,\cos\vartheta\,(1-2\sin^2\vartheta)$ we can compute I_4 as:

$$ I_4=4^{2p-1}\int_0^{\frac{\pi}{4}}\sin^{2p-1}\vartheta\,\cos^{2p-1}\vartheta\,(1-2\sin^2\vartheta)^{2p-1}\,\mathrm{d}\vartheta. $$

Now, changing the variable as $2\sin^2\vartheta=u$, we obtain

$$ I_4=4^{p-1}\int_0^1\frac{u^{p-1}(1-u)^{2p-1}}{(2-u)^{1-p}}\mathrm{d}u. $$

Let us collect the factor 2 in the denominator of I_4 to recognise its hypergeometric nature:

$$ I_4=8^{p-1}\int_0^1\frac{u^{p-1}(1-u)^{2p-1}}{(1-\frac{1}{2}u)^{1-p}}\mathrm{d}u=8^{p-1}\frac{\Gamma(p)\Gamma(2p)}{\Gamma(3p)}\,{}_2F_1\left(\begin{matrix}p;1-p\\ 3p\end{matrix}\,\middle|\,\frac{1}{2}\right). \tag{9.26} $$

Eventually (9.22b) follows comparing (9.25) and (9.26). □

Remark. There are many other evaluations of specific occurrences of the hypergeometric function, which we do not report here for the brevity of exposition. In any case, it is important to emphasise that this is more than just a mere exhibition of mathematical formalism in the sense that having (some) explicitly calculated values makes it possible to validate methods for the numerical approximation of these functions, which have great relevance in applications.

9.7 Variable transformations

We begin with an identity due to Pfaff known as Pfaff identity.

Theorem 9.6. If $\operatorname{Re} c > \operatorname{Re} b > 0$, $|x| < 1$ and $|x/(x-1)| < 1$ then

$$
{}_2\mathrm{F}_1\left(\begin{array}{c} a; b \\ c \end{array} \middle| \, x \right) = (1-x)^{-a} \, {}_2\mathrm{F}_1\left(\begin{array}{c} a; c-b \\ c \end{array} \middle| \, \frac{x}{x-1} \right). \tag{9.27}
$$

Proof. Change the variable in (9.8) as $u \to 1-u$ so that

$$
\begin{aligned}
{}_2\mathrm{F}_1\left(\begin{array}{c} a; b \\ c \end{array} \middle| \, x \right) &= \frac{\Gamma(c)}{\Gamma(b)\Gamma(c-b)} \int_0^1 \frac{u^{c-b-1}(1-u)^{b-1}}{(1-x+xu)^a} du \\
&= (1-x)^{-a} \frac{\Gamma(c)}{\Gamma(b)\Gamma(c-b)} \int_0^1 u^{c-b-1}(1-u)^{b-1} \left(1 - \frac{x}{x-1} u\right)^{-a} du \\
&= (1-x)^{-a} \, {}_2\mathrm{F}_1\left(\begin{array}{c} a; c-b \\ c \end{array} \middle| \, \frac{x}{x-1} \right).
\end{aligned}
$$

□

Remark. The proof we presented can be extended by complex analysis, which is well detailed in [74, page 247], and we report the conclusions made therein.

- Assumption $\operatorname{Re} c > \operatorname{Re} b > 0$, which is due to Theorem 9.3, can be relaxed using analytic continuation: as a matter of facts (9.27) holds with the sole restriction $c \neq 0, 1 - 1, -2, \ldots$

- Observing that

$$
\operatorname{Re} x < \frac{1}{2} \implies \left| \frac{x}{x-1} \right| < 1
$$

  we have obtained the analytic continuation of ${}_2\mathrm{F}_1$ in the half plane $\operatorname{Re} x < \frac{1}{2}$.

- Due to the symmetry property of the upper parameters of the hypergeometric series (9.5), we recognise a further identity:

$$
{}_2\mathrm{F}_1\left(\begin{array}{c} a; b \\ c \end{array} \middle| \, x \right) = (1-x)^{-b} \, {}_2\mathrm{F}_1\left(\begin{array}{c} b; c-a \\ c \end{array} \middle| \, \frac{x}{x-1} \right). \tag{9.27a}
$$

Due to Euler, the second identity follows from the iterated application of Pfaff identity (9.27).

Theorem 9.7. If $\operatorname{Re} c > \operatorname{Re} a > 0$ and $\operatorname{Re} c > \operatorname{Re} b > 0$ and if $|x/(x-1)| < 1$ then

$$ {}_2\mathrm{F}_1\left(\begin{array}{c} a;b \\ c \end{array} \middle| x\right) = (1-x)^{c-a-b}\,{}_2\mathrm{F}_1\left(\begin{array}{c} c-a;c-b \\ c \end{array} \middle| x\right). \tag{9.28} $$

Proof. Writing the right-hand side of the Pfaff identity (9.27) by exchanging the upper parameter and then applying again (9.27), which is indeed correct since we are assuming $|x/(x-1)| < 1$, we obtain

$$ \begin{aligned} {}_2\mathrm{F}_1\left(\begin{array}{c} c;b \\ c \end{array} \middle| x\right) &= (1-x)^{-a}\,{}_2\mathrm{F}_1\left(\begin{array}{c} a;c-b \\ c \end{array} \middle| \frac{x}{x-1}\right) \\ &= (1-x)^{-a}\,{}_2\mathrm{F}_1\left(\begin{array}{c} c-b;a \\ c \end{array} \middle| \frac{x}{x-1}\right) \\ &= (1-x)^{-a}\left(1-\frac{x}{x-1}\right)^{-(c-b)}\,{}_2\mathrm{F}_1\left(\begin{array}{c} c-b;c-a \\ c \end{array} \middle| \frac{\frac{x}{x-1}}{\frac{x}{x-1}-1}\right) \\ &= (1-x)^{c-b}\,{}_2\mathrm{F}_1\left(\begin{array}{c} c-b;c-a \\ c \end{array} \middle| x\right) \end{aligned} $$

ending the proof. □

We end the section by giving an example of a quadratic transformation of the hypergeometric function.

Theorem. If $|x| < 1$ and $a > 0$ then

$$ {}_2\mathrm{F}_1\left(\begin{array}{c} a;b \\ 2a \end{array} \middle| x\right) = \left(1-\frac{x}{2}\right)^{-b}\,{}_2\mathrm{F}_1\left(\begin{array}{c} \frac{b}{2};\frac{b+1}{2} \\ a+\frac{1}{2} \end{array} \middle| \left(\frac{x}{2-x}\right)^2\right). \tag{9.29} $$

Proof. We follow the demonstration in [5, page 127] but provide the details of many omitted steps. First, we observe that both series in (9.29) converge for $-1 < x < 1$, then we use (9.8), permissible since $a > 0$ and operate some algebraic transformations given a change of variable:

$$ \begin{aligned} {}_2\mathrm{F}_1\left(\begin{array}{c} a;b \\ 2a \end{array} \middle| x\right) &= \frac{\Gamma(2a)}{\Gamma^2(a)}\int_0^1 \frac{\left(s-s^2\right)^{a-1}}{(1-xs)^b}\,\mathrm{d}s \\ &= \frac{\Gamma(2a)}{\Gamma^2(a)}\int_0^1 \frac{\left[\left(\frac{1}{2}\right)^2-\left(\frac{1}{2}-s\right)^2\right]^{a-1}}{(1-xs)^b}\,\mathrm{d}s \\ &= \frac{\Gamma(2a)}{\Gamma^2(a)}\left(\frac{1}{2}\right)^{2(a-1)}\int_0^1 \frac{\left[1-(1-2s)^2\right]^{a-1}}{(1-xs)^b}\,\mathrm{d}s. \end{aligned} $$

We change variables in the last integral by placing $1-2s=t$ so that

$$ {}_2F_1\left(\begin{array}{c} a;b \\ 2a \end{array}\middle|\, x\right) = \frac{\Gamma(2a)}{\Gamma^2(a)}\left(\frac{1}{2}\right)^{2a-1}\int_{-1}^{1}\frac{(1-t^2)^{a-1}}{\left[1-\frac{x}{2}(1-t)\right]^b}\,\mathrm{d}t. $$

For the right-hand side integral, we need to "conveniently prepare" the denominator using the following equalities:

$$ \left[1-\frac{x}{2}(1-t)\right]^b = \left(1-\frac{x}{2}\right)^b\left[1+\frac{\frac{x}{2}}{1-\frac{x}{2}}\,t\right]^b = \left(1-\frac{x}{2}\right)^b\left[1-\frac{x}{x-2}\,t\right]^b. $$

Hence

$$ {}_2F_1\left(\begin{array}{c} a;b \\ 2a \end{array}\middle|\, x\right) = \frac{\Gamma(2a)}{\Gamma^2(a)}\left(\frac{1}{2}\right)^{2a-1}\frac{2^b}{(2-x)^b}\int_{-1}^{1}\frac{(1-t^2)^{a-1}}{\left(1-\frac{x}{x-2}\,t\right)^b}\,\mathrm{d}t. \tag{9.30} $$

Now, in (9.30), we express the denominator in the binomial form:

$$ \left(1-\frac{x}{x-2}\,t\right)^{-b} = \sum_{n=0}^{\infty}\frac{(b)_n}{n!}\left(\frac{x}{x-2}\,t\right)^n. $$

So that

$$ {}_2F_1\left(\begin{array}{c} a;b \\ 2a \end{array}\middle|\, x\right) = $$

$$ \frac{\Gamma(2a)}{\Gamma^2(a)}\left(\frac{1}{2}\right)^{2a-1}\frac{1}{(1-\frac{x}{2})^b}\sum_{n=0}^{\infty}\frac{(b)_n}{n!}\left(\frac{x}{x-2}\right)^n\int_{-1}^{1}t^n\,(1-t^2)^{a-1}\,\mathrm{d}t. \tag{9.31} $$

We observe that in the series in (9.31), integrals of odd index vanish; this equality is therefore rewritten as, also taking into account the symmetry given by the integration of an even function

$$ {}_2F_1\left(\begin{array}{c} a;b \\ 2a \end{array}\middle|\, x\right) = $$

$$ \frac{\Gamma(2a)}{\Gamma^2(a)}\left(\frac{1}{2}\right)^{2a-1}\frac{1}{(1-\frac{x}{2})^b}\sum_{m=0}^{\infty}\frac{(b)_{2m}}{(2m)!}\left(\frac{x}{x-2}\right)^{2m}2\int_{0}^{1}t^{2m}\,(1-t^2)^{a-1}\,\mathrm{d}t. \tag{9.32} $$

Finally, the integral in (9.32) can be evaluated in terms of the beta function:

$$ 2\int_0^1 t^{2m}\,(1-t^2)^{a-1}\,\mathrm{d}t = \int_0^1(1-t)^{a-1}t^{m-\frac{1}{2}}\,\mathrm{d}t = \frac{\Gamma(a)\Gamma\left(m+\frac{1}{2}\right)}{\Gamma\left(a+m+\frac{1}{2}\right)}. $$

Thus, going back to (9.32) we arrive at

$$
{}_2\mathrm{F}_1\left(\begin{array}{c} a;b \\ 2a \end{array}\middle| x\right) =
$$

$$
\frac{2^{1-2a}\,\Gamma(2a)}{\Gamma(a)}\frac{1}{(1-\frac{x}{2})^b}\sum_{m=0}^{\infty}\frac{(b)_{2m}}{(2m)!}\frac{\Gamma\left(m+\frac{1}{2}\right)}{\Gamma\left(a+m+\frac{1}{2}\right)}\left[\left(\frac{x}{x-2}\right)^2\right]^m . \qquad (9.33)
$$

At this point, theorem 2.6 comes into play:

$$
\frac{(b)_{2m}}{(2m)!}=\frac{(b)_{2m}}{(1)_{2m}}=\frac{2^{2m}\left(\frac{b}{2}\right)_m\left(\frac{b}{2}+\frac{1}{2}\right)_m}{2^{2m}\left(\frac{1}{2}\right)_m\left(\frac{1}{2}+\frac{1}{2}\right)_m}=\frac{\left(\frac{b}{2}\right)_m\left(\frac{b}{2}+\frac{1}{2}\right)_m}{m!\left(\frac{1}{2}\right)_m}.
$$

Then, it observed that

$$
\left(\tfrac{1}{2}\right)_m=\frac{\Gamma\left(m+\frac{1}{2}\right)}{\Gamma(\frac{1}{2})},\quad \Gamma\left(a+m+\frac{1}{2}\right)=\Gamma\left(a+\tfrac{1}{2}\right)\left(a+\tfrac{1}{2}\right)_m .
$$

By inserting in(9.33) we finally get

$$
{}_2\mathrm{F}_1\left(\begin{array}{c} a;b \\ 2a \end{array}\middle| x\right)=\frac{2^{1-2a}\,\Gamma(2a)\sqrt{\pi}}{\Gamma(a)\Gamma(a+\frac{1}{2})}\frac{1}{(1-\frac{x}{2})^b}\sum_{m=0}^{\infty}\frac{\left(\frac{b}{2}\right)_m\left(\frac{b}{2}+\frac{1}{2}\right)_m}{\left(a+\frac{1}{2}\right)_m\,m!}\left[\left(\frac{x}{x-2}\right)^2\right]^m .
$$

Formula (9.29) follows from Legendre duplication formula for gamma (2.50) from which we derive:

$$
\frac{2^{1-2a}\,\Gamma(2a)\sqrt{\pi}}{\Gamma(a)\Gamma(a+\frac{1}{2})}=1.
$$

□

Chapter 10

Generalised hypergeometric functions of one variable

In this chapter, we investigate the "natural" generalisation of the hypergeometric series, introducing, roughly speaking, more parameters, restricting ourselves, however, in this chapter to considering functions of only one variable. This is not an instrumental exercise because the functions we obtain from this generalisation have innumerable fields of use. This chapter attempts to present a vast scientific literature on this subject, referring for further study, for instance, to [80, 96, 105].

10.1 The ${}_pF_q$ power series

A generalised hypergeometric series is a power series in which the ratio of two consecutive coefficients to the summation index n is a rational function of n. Such a series, if convergent, defines a generalised hypergeometric function, which may be continued to a larger domain. We will deal with generalised hypergeometric series of the Gaussian type, motivated by the structure of the hypergeometric series (9.5). The generalised Gaussian hypergeometric function defined by the power series

$${}_pF_q\left(\begin{array}{c} a_1, \dots, a_p \\ b_1, \dots, b_q \end{array} \middle| x \right) = \sum_{n=0}^{\infty} \frac{(a_1)_n \cdots (a_p)_n}{(b_1)_n \cdots (b_q)_n} \frac{x^n}{n!}. \tag{10.1}$$

Here $(a)_n$ is, as usual, a Pochhammer symbol, while $a_1, \dots, a_p$ and $b_1, \dots, b_q$ real or complex parameters and p, q positive integers. In this situation, as well as for the ${}_2F_1$ case, if any of the parameters at numerator a_i in (10.1) is either zero or a negative integer, the series reduces to a polynomial. Generalised hypergeometric functions include numerous elementary and non-elementary functions as special cases. For example, we find the exponential function if there are no parameters in the numerator and denominator:

$${}_0F_0\left(\begin{array}{c} - \\ - \end{array} \middle| x \right) = \sum_{k=0}^{\infty} \frac{x^k}{k!} = e^x.$$

DOI: 10.1201/9781003171546-10

The binomial series can also be seen as a hypergeometric series:

$$ {}_1F_0\left(\begin{matrix} a \\ - \end{matrix}\,\middle|\, x\right) = \sum_{k=0}^{\infty} \frac{(a)_k}{k!} x^k = (1-x)^{-a}, \quad |x| < 1. $$

The goniometric and hyperbolic functions are hypergeometric series:

$$ \begin{aligned} \sin x &= \sum_{n=0}^{\infty} \frac{(-1)^n}{(2n+1)!} x^{2n+1} = x\, {}_0F_1\left(\begin{matrix} - \\ \frac{3}{2} \end{matrix}\,\middle|\, -\frac{x^2}{4}\right), \\ \cos x &= \sum_{n=0}^{\infty} \frac{(-1)^n}{(2n)!} x^{2n} = {}_0F_1\left(\begin{matrix} - \\ \frac{1}{2} \end{matrix}\,\middle|\, -\frac{x^2}{4}\right), \\ \sinh x &= \sum_{n=0}^{\infty} \frac{1}{(2n+1)!} x^{2n+1} = x\, {}_0F_1\left(\begin{matrix} - \\ \frac{3}{2} \end{matrix}\,\middle|\, \frac{x^2}{4}\right), \\ \cosh x &= \sum_{n=0}^{\infty} \frac{1}{(2n)!} x^{2n} = {}_0F_1\left(\begin{matrix} - \\ \frac{1}{2} \end{matrix}\,\middle|\, \frac{x^2}{4}\right). \end{aligned} $$

The logarithmic, dilogarithmic and trilogarithm function $\mathrm{Li}_2(x)$ and $\mathrm{Li}_3(x)$ are defined for $|x| < 1$ by

$$ \begin{aligned} \mathrm{Li}_1(x) &= \sum_{n=1}^{\infty} \frac{x^n}{n} = x\, {}_2F_1\left(\begin{matrix} 1,1 \\ 2 \end{matrix}\,\middle|\, x\right), \\ \mathrm{Li}_2(x) &= \sum_{n=1}^{\infty} \frac{x^n}{n^2} = x\, {}_3F_2\left(\begin{matrix} 1,1,1 \\ 2,2 \end{matrix}\,\middle|\, x\right), \\ \mathrm{Li}_3(x) &= \sum_{n=1}^{\infty} \frac{x^n}{n^3} = x\, {}_4F_3\left(\begin{matrix} 1,1,1,1 \\ 2,2,2 \end{matrix}\,\middle|\, x\right). \end{aligned} $$

The reason for the name given to these functions follows from $\mathrm{Li}_1(x) = -\ln(1-x)$.

The behaviour of Li_1, Li_2, and Li_3 makes us realise that when a generalised hypergeometric series is assigned, its radius of convergence must be determined: as a matter of fact from the ratio test we can state:

Theorem. For the generalised hypergeometric series (10.1), the following statements hold:

- If $p > q + 1$ (10.1) diverges for every $x \neq 0$.
- If $p = q + 1$ (10.1) converges for $|x| < 1$ and diverges for $|x| > 1$.
- If $p \leq q$, (10.1) converges for each x.

For the proof see [96, page 73–74].

10.2 Hypergeometric differential equation

In the case where $p = q + 1$, if we define the differential operator $\delta := x\,\mathrm{d}/\mathrm{d}x$, then the function ${}_{q+1}\mathrm{F}_q$ solves the differential equation:

$$\delta\,(\delta + b_1 - 1)\cdots(\delta + b_q - 1)\,y = x\,(\delta + a_1)\cdots(\delta + a_p)\,y. \tag{10.2}$$

The general solution to (10.2) is described by the functions:

$$x^{1-b_\nu}\,{}_{q+1}\mathrm{F}_q\left(\begin{matrix} 1 + a_1 - b_\nu, \dots, 1 + a_{q+1} - b_\nu \\ 1 + b_1 - b_\nu, \dots * \dots, 1 + b_q - b_\nu, 2 - b_\nu \end{matrix} \,\middle|\, x\right),$$

for $\nu = 1, \dots, q$ and where $*$ indicates that $1 + b_\nu - b_\nu$ is omitted from the sequence of denominators.

Example. The hypergeometric function

$${}_3\mathrm{F}_2\left(\begin{matrix} a_1, a_2, a_3 \\ b_1, b_2 \end{matrix} \,\middle|\, x\right)$$

solves the third-order linear differential equation:

$$\begin{aligned}&(1 - x)x^2 y^{(3)} - x\,[x(a_1 + a_2 + a_3 + 3) - b_1 - b_2 - 1]\,y'' + \\ &[b_1 b_2 - x(a_1 a_2 + a_1 a_3 + a_2 a_3 + a_1 + a_2 + a_3 + 1)]\,y' - a_1 a_2 a_3 y = 0.\end{aligned}$$

The general solution is

$$\begin{aligned} y(x) &= c_1\,{}_3\mathrm{F}_2\left(\begin{matrix} a_1, a_2, a_3 \\ b_1, b_2 \end{matrix} \,\middle|\, x\right) \\ &\quad + c_2\,x^{1-b_1}\,{}_3\mathrm{F}_2\left(\begin{matrix} a_1 - b_1 + 1, a_2 - b_1 + 1, a_3 - b_1 + 1 \\ 2 - b_1, b_2 - b_1 + 1 \end{matrix} \,\middle|\, x\right) \\ &\quad + c_3\,x^{1-b_2}\,{}_3\mathrm{F}_2\left(\begin{matrix} a_1 - b_2 + 1, a_2 - b_2 + 1, a_3 - b_2 + 1 \\ 2 - b_2, b_1 - b_2 + 1 \end{matrix} \,\middle|\, x\right). \end{aligned}$$

10.3 Integral representation theorem

The integral representation Theorem 9.3 extends to generalised hypergeometric functions.

Theorem. If $p \leq q+1$ and $b_1 > a_1 > 0$ and if none of the $b_1, b_2, \dots, b_q$ is zero or a non-negative integer, then

$$
{}_p\mathrm{F}_q\left(\begin{array}{c} a_1,\dots,a_p \\ b_1,\,\dots,b_q \end{array}\middle|\, x\right)
= \frac{\Gamma(b_1)}{\Gamma(a_1)\Gamma(b_1-a_1)}\int_0^1 t^{a_1-1}(1-t)^{b_1-a_1-1}\,{}_{p-1}\mathrm{F}_{q-1}\left(\begin{array}{c} a_2,\dots,a_p \\ b_2,\,\dots,b_q \end{array}\middle|\, t\,x\right)\mathrm{d}t. \tag{10.3}
$$

Proof. The assumptions on the coefficients ensure the convergence of the integral on the right-hand side of (10.3). Denoted by I this integral, the generalised hypergeometric series is made explicit, and noting that under the stated hypotheses, we can exchange between the series and the integral, we obtain the thesis; here are the details:

$$
\begin{aligned}
I &= \int_0^1 t^{a_1-1}(1-t)^{b_1-a_1-1}\sum_{n=0}^{\infty}\frac{(a_2)_n\cdots(a_p)_n}{(b_2)_n\cdots(b_q)_n}\frac{(xt)^n}{n!}\mathrm{d}t \\
&= \sum_{n=0}^{\infty}\frac{(a_2)_n\cdots(a_p)_n}{(b_2)_n\cdots(b_q)_n}\frac{x^n}{n!}\int_0^1 t^{a_1+n-1}(1-t)^{b_1-a_1-1}\mathrm{d}t \\
&= \sum_{n=0}^{\infty}\frac{(a_2)_n\cdots(a_p)_n}{(b_2)_n\cdots(b_q)_n}\frac{x^n}{n!}\mathrm{B}(a_1+n, b_1-a_1) \\
&= \sum_{n=0}^{\infty}\frac{(a_2)_n\cdots(a_p)_n}{(b_2)_n\cdots(b_q)_n}\frac{x^n}{n!}\frac{\Gamma(a_1+n)\Gamma(b_1-a_1)}{\Gamma(b_1+n)} \\
&= \sum_{n=0}^{\infty}\frac{(a_2)_n\cdots(a_p)_n}{(b_2)_n\cdots(b_q)_n}\frac{x^n}{n!}\frac{(a_1)_n\Gamma(a_1)\Gamma(b_1-a_1)}{(b_1)_n\Gamma(b_1)} \\
&= \sum_{n=0}^{\infty}\frac{(a_1)_n(a_2)_n\cdots(a_p)_n}{(b_1)_n(b_2)_n\cdots(b_q)_n}\frac{x^n}{n!}\frac{\Gamma(a_1)\Gamma(b_1-a_1)}{\Gamma(b_1)} \\
&= {}_p\mathrm{F}_q\left(\begin{array}{c} a_1,\dots,a_p \\ b_1,\,\dots,b_q \end{array}\middle|\, x\right)\frac{\Gamma(a_1)\Gamma(b_1-a_1)}{\Gamma(b_1)}.
\end{aligned}
$$

□

10.4 Incomplete Gamma functions

The gamma function was introduced using an integral defined on the positive semi-axis. In applications, it is useful to consider the same integral both in closed and limited intervals of the type $[0, x]$ and in the complementary intervals $(x, +\infty)$: in this way, we arrive at defining the two incomplete gamma functions, which, as we shall see, have a hypergeometric nature.

Definition. If $x > 0$ e $\mathrm{Re}\, a > 0$ the integrals

$$\gamma(a,x) = \int_0^x e^{-t} t^{a-1} \mathrm{d}t \quad (10.4a) \qquad \Gamma(a,x) = \int_x^\infty e^{-t} t^{a-1} \mathrm{d}t \quad (10.4b)$$

are called the lower incomplete gamma function and upper incomplete gamma function.

The following equality is trivial:

$$\Gamma(x) = \gamma(a,x) + \Gamma(a,x)$$

The hypergeometric nature of $\gamma(a,x)$ is highlighted by the following theorem.

Theorem.

$$\gamma(a,x) = \frac{x^a}{a}\, {}_1\mathrm{F}_1\left(\begin{array}{c} a \\ a+1 \end{array} \middle| -x \right) \tag{10.5}$$

Proof. We develop the exponential in power series:

$$\begin{aligned} \gamma(a,x) &= \int_0^x t^{a-1} \sum_{n=0}^\infty \frac{(-1)^n}{n!} t^n \mathrm{d}t = \sum_{n=0}^\infty \frac{(-1)^n}{n!} \int_0^x t^{n+a-1} \mathrm{d}t \\ &= \sum_{n=0}^\infty \frac{(-1)^n}{n!} \left[\frac{t^{n+a}}{n+a} \right]_{t=0}^{t=x} = x^a \sum_{n=0}^\infty \frac{(-1)^n}{n!} \frac{x^n}{n+a}. \end{aligned}$$

Now, recalled the property (2.13) of the Pochhammer symbol, we have that

$$\frac{(a)_n}{(a+1)_n} = \frac{\Gamma(a+n)}{\Gamma(a)} \frac{\Gamma(a+1)}{\Gamma(a+1+n)} = \frac{\Gamma(a+n)}{\Gamma(a)} \frac{a\Gamma(a)}{(a+n)\Gamma(a+n)} = \frac{a}{a+n}.$$

Hence

$$\gamma(a,x) = \frac{x^a}{a} \sum_{n=0}^\infty \frac{(a)_n}{(a+1)_n} \frac{(-x)^n}{n!} = \frac{x^a}{a}\, {}_1\mathrm{F}_1\left(\begin{array}{c} a \\ a+1 \end{array} \middle| -x \right),$$

showing (10.5) and, thus, ending the proof. □

It is natural to refer to these functions as incomplete Eulerian functions. They are frequently used in mathematical statistics.

10.5 Kummer function

The ${}_1\mathrm{F}_1$ function is said Kummer function. It is also known as the confluent hypergeometric function of the first kind and solves the ordinary differential equation:

$$x\,u'' + (b-x)\,u' - a\,u = 0. \tag{10.6}$$

This function is also represented by the letter M. We notice that the power series

$$M\left(\begin{array}{c} a \\ b \end{array} \middle| x\right) = {}_1\mathrm{F}_1\left(\begin{array}{c} a \\ b \end{array} \middle| x\right) = \sum_{n=0}^{\infty} \frac{(a)_n}{(b)_n} \frac{x^n}{n!} \tag{10.7}$$

has an infinite radius of convergence, as can be seen immediately using the ratio test: in fact, the ratio between two consecutive terms of (10.7)

$$\frac{x_{n+1}}{x_n} = \frac{(a)_{n+1}}{(b)_{n+1}} \frac{x^{n+1}}{(n+1)!} \frac{(b)_n}{(a)_n} \frac{n!}{x^n} = \frac{x(a+n)}{(n+1)(b+n)},$$

vanishes for $n \to \infty$ for any $x \in \mathbb{C}$.

The term confluent is motivated by the next proposition, which illustrates that $M(a, b; x)$ can be understood as a limit case of ${}_2\mathrm{F}_1$.

Theorem. For any a, $c \in \mathbb{C}$

$${}_1\mathrm{F}_1\left(\begin{array}{c} a \\ c \end{array} \middle| x\right) = \lim_{b \to \infty} {}_2\mathrm{F}_1\left(\begin{array}{c} a, b \\ c \end{array} \middle| \frac{x}{b}\right). \tag{10.8}$$

Proof. Recalling the definition of ${}_2\mathrm{F}_1$, which we compute for x/b we have

$${}_2\mathrm{F}_1\left(\begin{array}{c} a; b \\ c \end{array} \middle| \frac{x}{b}\right) = \sum_{n=0}^{+\infty} \frac{(a)_n (b)_n}{(c)_n \, n!} \left(\frac{x}{b}\right)^n.$$

Formula (10.8) follows from

$$\frac{(b)_n}{b^n} = 1 \cdot \left(1 + \frac{1}{b}\right) \cdot \left(1 + \frac{2}{b}\right) \cdots \cdot \left(1 + \frac{n-1}{b}\right).$$

□

An Eulerian integral representation theorem also holds for the Kummer function.

Theorem. If $b > a > 0$ then

$${}_1\mathrm{F}_1\left(\begin{array}{c} a \\ b \end{array} \middle| x\right) = \frac{\Gamma(b)}{\Gamma(a)\Gamma(b-a)} \int_0^1 e^{tx} t^{a-1} (1-t)^{b-a-1} \mathrm{d}t. \tag{10.9}$$

Proof. We calculate the integral on the left-hand side of (10.9). We expand the exponential in power series and, then, exchange the series with the integral (monotone convergence theorem)

$$\begin{aligned} \int_0^1 e^{tx} t^{a-1} (1-t)^{b-a-1} \mathrm{d}t &= \int_0^1 \sum_{n=0}^{\infty} \frac{x^n t^n}{n!} t^{a-1} (1-t)^{b-a-1} \mathrm{d}t \\ &= \sum_{n=0}^{\infty} \frac{x^n}{n!} \int_0^1 t^{n+a-1} (1-t)^{b-a-1} \mathrm{d}t = \sum_{n=0}^{\infty} \frac{x^n}{n!} \mathrm{B}(a+n, b-a) \\ &= \sum_{n=0}^{\infty} \frac{x^n}{n!} \frac{\Gamma(a+n)\Gamma(b-a)}{\Gamma(b+n)}. \end{aligned}$$

Using $\Gamma(a+n) = (a)_n\Gamma(a)$ and $\Gamma(b+n) = (b)_n\Gamma(b)$ we arrive at

$$\int_0^1 e^{tx}t^{a-1}(1-t)^{b-a-1}\mathrm{d}t = \sum_{n=0}^{\infty} \frac{x^n}{n!}\frac{(a)_n\Gamma(a)}{(b)_n\Gamma(b)}\Gamma(b-a)$$
$$= \frac{\Gamma(a)\Gamma(b-a)}{\Gamma(b)}\sum_{n=0}^{\infty}\frac{x^n}{n!}\frac{(a)_n}{(b)_n}.$$

Hence, (10.9) follows. □

The general solution of (10.6) is completed by the confluent hypergeometric function of the second kind, Tricomi's function:

$$U\left(\begin{array}{c} a \\ b \end{array}\middle| x\right) = \frac{\pi}{\sin(\pi b)}\left(\frac{1}{\Gamma(b)\Gamma(a-b+1)}\,{}_1\mathrm{F}_1\left(\begin{array}{c} a \\ b \end{array}\middle| x\right) - \frac{x^{1-b}}{\Gamma(2-b)\Gamma(a)}\,{}_1\mathrm{F}_1\left(\begin{array}{c} a-b+1 \\ 2-b \end{array}\middle| x\right)\right). \tag{10.10}$$

This function has the integral representation, valid for $a > 0$, $x > 0$

$$U\left(\begin{array}{c} a \\ b \end{array}\middle| x\right) = \frac{1}{\Gamma(a)}\int_0^{\infty} e^{-xt}t^{a-1}(1+t)^{b-a+1}\mathrm{d}t. \tag{10.11}$$

We do not demonstrate here that (10.10) completes the general solution of equation (10.6), and we leave to the reader the proof of (10.11).

10.6 The error function

The error function, in terms of the probability integral, is defined as:

$$\mathrm{erf}(x) = \frac{2}{\sqrt{\pi}}\int_0^x e^{-t^2}\mathrm{d}t. \tag{10.12}$$

Notice that (10.12) can be represented in terms of the Kummer function.

Theorem. For any $x \in \mathbb{R}$ we have

$$\mathrm{erf}(x) = \frac{2x}{\sqrt{\pi}}\,{}_1\mathrm{F}_1\left(\begin{array}{c} \frac{1}{2} \\ \frac{3}{2} \end{array}\middle| -x^2\right). \tag{10.13}$$

Proof. To show (10.13) we expand the exponential e^{-t^2} and then integrate term by term arriving at

$$\int_0^x e^{-t^2}\mathrm{d}t = \int_0^x \sum_{n=0}^{\infty}\frac{(-1)^n}{n!}t^{2n}\mathrm{d}t = x\sum_{n=0}^{\infty}\frac{(-x^2)^n}{(2n+1)n!}.$$

The formula (10.13) follows from the identity:

$$\frac{1}{2n+1} = \frac{\left(\frac{1}{2}\right)_n}{\left(\frac{3}{2}\right)_n}.$$

□

The error function is related to the incomplete gamma function by the following relation.

$$\gamma\left(\frac{1}{2}, 1\right) = \sqrt{\pi}\,\mathrm{erf}(1). \tag{10.14}$$

In fact, since

$$\gamma\left(\frac{1}{2}, 1\right) = \int_0^1 t^{\frac{1}{2}-1} e^{-t}\,\mathrm{d}t,$$

the change of variable $x^2 = t$ provides (10.14).

Chapter 11

Applications and exercises

This final chapter illustrates some applications of the special functions exposed up to now, starting from classic evaluations of defined integrals, series summation, exact integration of differential equations, the solution of algebraic equations of degree higher than the fourth, and probabilistic applications. In the initial examples on the calculation of integrals and series summation, I also drew on many amateur contents, which have been populating the social networks: making an exhaustive list of sources is as difficult as it is probably useless. The few exercises we have chosen to present here are certainly not original.

For the reader's benefit, we will indicate in which chapter the requirements for understanding the resolution procedure are found after each statement. This way, this chapter can be read piecewise.

11.1 Series summations

The relationship between beta and the binomial coefficients allows the summation of several in an infinite series of great interest and is also relevant for the numerical evaluation of π. The following summations 1, 2, and 4 are all derived using representations of the central binomial coefficients; for 5 and 3, the beta function also plays a fundamental role in the summation process. Summation 8 requires more advanced methods in hypergeometric functions.

Summation 1.

$$\sum_{n=0}^{\infty} \frac{1}{\dbinom{2n}{n}} = \frac{4}{3} + \frac{2\pi}{9\sqrt{3}}. \tag{11.1}$$

Proof. Recall identity (2.29) at page 69, which here yields:

$$\begin{aligned}\sum_{n=0}^{\infty} \frac{1}{\dbinom{2n}{n}} &= \sum_{n=0}^{\infty}(2n+1)\int_0^1 x^n(1-x)^n \mathrm{d}x \\ &= \int_0^1 \sum_{n=0}^{\infty}(2n+1)\left(x(1-x)\right)^n \mathrm{d}x. \qquad (*)\end{aligned}$$

DOI: 10.1201/9781003171546-11

The exchange between series and integral is correct since it is a series with positive terms. The series that appears in the integrand in $(*)$ is expressed in closed form as a consequence of the formula for the sum of the geometric series; write for ease $u = x(1-x)$ and, first recall the formula, which holds for $-1 < u < 1$

$$\sum_{n=0}^{\infty} u^n = \frac{1}{1-u}$$

and then invoke Theorem 1.7, which enables the differentiation term by term yielding:

$$\sum_{n=0}^{\infty} n u^{n-1} = \frac{1}{(1-u)^2} = \sum_{n=1}^{\infty} (n+1) u^n .$$

Thus, we infer, recalling that we are considering u values such that the infinite series is convergent

$$\begin{aligned}
\sum_{n=1}^{\infty} (2n+1) u^n &= \sum_{n=1}^{\infty} (n+1) u^n + \sum_{n=1}^{\infty} n u^n \\
&= \sum_{n=1}^{\infty} (n+1) u^n + \frac{1}{u} \sum_{n=1}^{\infty} n u^{n-1} \\
&= \frac{1}{(1-u)^2} + u \frac{1}{(1-u)^2} \qquad (\dagger) \\
&= \frac{1+u}{(1-u)^2} .
\end{aligned}$$

Going back to x, we have the summation formula, valid for $0 < x < 1$:

$$\sum_{n=0}^{\infty} (2n+1) \left(x(1-x) \right)^n = \frac{1+x-x^2}{\left(1-x+x^2\right)^2} -$$

By inserting this sum formula into $(*)$, we arrive at, without going into details regarding the (undergraduate) passages to arrive at $(**)$:

$$\begin{aligned}
\sum_{n=0}^{\infty} \frac{1}{\binom{2n}{n}} &= \int_0^1 \frac{1+x-x^2}{\left(1-x+x^2\right)^2} \mathrm{d}x \\
&= \left[\frac{2(2x-1)}{3\,(1+x-x^2)} + \frac{2}{3\sqrt{3}} \arctan\left(\frac{2x-1}{\sqrt{3}} \right) \right]_0^1 \qquad (**) \\
&= \frac{2}{3} + \frac{\pi}{9\sqrt{3}} - \left(-\frac{2}{3} - \frac{\pi}{9\sqrt{3}} \right) ,
\end{aligned}$$

which provides (11.1). □

Summation 2.

$$\sum_{n=1}^{\infty} \frac{1}{n\binom{2n}{n}} = \frac{\pi}{3\sqrt{3}}. \tag{11.2}$$

Proof. Prerequisites chapter 2. Expanding the central binomial coefficient on the left-hand side of (11.2) and carrying out the calculations directly, we find

$$\begin{aligned}
\sum_{n=1}^{\infty} \frac{1}{n\binom{2n}{n}} &= \sum_{n=1}^{\infty} \frac{(n!)^2}{n\,(2n)!} = \sum_{n=1}^{\infty} \frac{(n!)^2}{2n^2\,(2n-1)!} \\
&= \frac{1}{2}\sum_{n=1}^{\infty} \frac{(n-1)!\,(n-1)!}{(2n-1)!} = \frac{1}{2}\sum_{n=1}^{\infty} \frac{\Gamma(n)\,\Gamma(n)}{\Gamma(2n)} \\
&= \frac{1}{2}\sum_{n=1}^{\infty} \mathrm{B}(n,n) \\
&= \frac{1}{2}\sum_{n=1}^{\infty} \int_0^1 x^{n-1}(1-x)^{n-1}.
\end{aligned}$$

We once again use the summation formula of the geometric series here; the common ratio is $x(1-x)$; we leave the task of sorting out the details to the reader, obtaining the transformation of the series into a definite integral, arriving at

$$\sum_{n=1}^{\infty} \frac{1}{n\binom{2n}{n}} = \frac{1}{2}\int_0^1 \frac{\mathrm{d}x}{1-x+x^2}.$$

The formula (11.2) eventually follows from the evaluation of the last (elementary) integral on the right-hand side. □

The following summation is similar to the previous, but we present it in detail since the slight modification in the general term of the series has the result that its sum is indeed π.

Summation 3.

$$\sum_{n=0}^{\infty} \frac{2^{n+1}(n!)^2}{(2n+1)!} = \pi. \tag{11.3}$$

Proof. Prerequisites chapter 2. For $n \in \mathbb{N}$ we have

$$\mathrm{B}(n+1,n+1) = \frac{\Gamma^2(n+1)}{\Gamma(2n+2)} = \frac{(n!)^2}{(2n+1)!}. \tag{11.4}$$

On the other hand, we also have

$$\mathrm{B}(n+1,n+1) = \int_0^1 x^n(1-x)^n \mathrm{d}x. \tag{11.5}$$

We sum the series to the left-hand side of (11.3), exploiting the relations (11.4) and (11.5):

$$\sum_{n=0}^{\infty} \frac{2^{n+1}(n!)^2}{(2n+1)!} = \sum_{n=0}^{\infty} 2^{n+1} \mathrm{B}(n+1, n+1) = \sum_{n=0}^{\infty} 2^{n+1} \int_0^1 x^n (1-x)^n \mathrm{d}x.$$

The exchange between integral and series is possible due to the positivity of the integrand sequence; therefore

$$\sum_{n=0}^{\infty} \frac{2^{n+1}(n!)^2}{(2n+1)!} = \int_0^1 \sum_{n=0}^{\infty} 2^{n+1} x^n (1-x)^n \mathrm{d}x = \int_0^1 \frac{2}{2x^2 - 2x + 1} \mathrm{d}x.$$

The summation of the series is then reduced to the calculation of an elementary integral. Changing variable $x = \frac{1}{2}y + \frac{1}{2}$ we obtain

$$\int_0^1 \frac{2}{2x^2 - 2x + 1} \mathrm{d}x = \int_0^1 \frac{2}{x^2 + (x-1)^2} \mathrm{d}x = \int_{-1}^1 \frac{2}{1+y^2} \mathrm{d}y = \pi.$$

So, our claim is proven. □

In the following summation, we merge the representation of the central binomial coefficient with the Legendre duplication formula for gamma (2.50).

Summation 4.

$$\sum_{n=1}^{\infty} \frac{1}{n^2 \binom{2n}{n}} = \frac{\pi^2}{18}. \tag{11.6}$$

Proof. Prerequisites chapter 2. Notice that, for $\Gamma(n+1) = n!$ and for $n \in \mathbb{N}$ we have

$$\binom{2n}{n} = \frac{(2n)!}{n!\, n!} = \frac{2\,\Gamma(2n)}{n \Gamma^2(n)}.$$

Now we insert the representation of $\Gamma(2n)$ given in Theorem 2.15, i.e. the Legendre duplication formula (2.50), obtaining

$$\binom{2n}{n} = \frac{2^{2n} \Gamma\left(n + \frac{1}{2}\right)}{n\, \Gamma\left(\frac{1}{2}\right) \Gamma(n)}.$$

We can, therefore, use this fact in the series (11.6), obtaining

$$\frac{1}{n^2 \binom{2n}{n}} = \frac{\Gamma\left(\frac{1}{2}\right) \Gamma(n)}{2^{2n}\, n\, \Gamma\left(n + \frac{1}{2}\right)} = \frac{1}{n\, 2^{2n}} \mathrm{B}\left(n, \frac{1}{2}\right).$$

Notice that the last step follows from Theorem 2.9. Now, by definition, see (2.25), we have

$$\mathrm{B}\left(n, \frac{1}{2}\right) = \int_0^1 \frac{x^{n-1}}{\sqrt{1-x}} \mathrm{d}x.$$

Going back to (11.6), exchanging between the series and the integral, we arrive at

$$\sum_{n=1}^{\infty} \frac{1}{n^2 \binom{2n}{n}} = \int_0^1 \frac{1}{\sqrt{1-x}} \sum_{n=1}^{\infty} \frac{x^{n-1}}{n\, 2^{2n}}\, \mathrm{d}x = \int_0^1 \frac{1}{x\sqrt{1-x}} \sum_{n=1}^{\infty} \frac{1}{n} \left(\frac{x}{4}\right)^n \mathrm{d}x.$$

Recalling Theorem 1.15 (logarithmic series) page 11:

$$\sum_{n=1}^{\infty} \frac{1}{n} \left(\frac{x}{4}\right)^n = -\ln\left(\frac{4-x}{4}\right),$$

we can transform series (11.6) into a definite integral

$$\sum_{n=1}^{\infty} \frac{1}{n^2 \binom{2n}{n}} = -\int_0^1 \frac{1}{x\sqrt{1-x}} \ln\left(1-\frac{x}{4}\right) \mathrm{d}x. \tag{11.7}$$

Given that the integral (11.7) is not easy to evaluate, we use a technique that we adopted in the proof of Theorem 3.9. We observe that

$$-\frac{1}{x} \ln\left(1-\frac{x}{4}\right) = \int_0^{\frac{1}{4}} \frac{1}{1-x\,y}\, \mathrm{d}y.$$

Thus, inserting in (11.7), we can represent the one-dimensional integral as a double integral, arriving at

$$\sum_{n=1}^{\infty} \frac{1}{n^2 \binom{2n}{n}} = \int_0^1 \frac{1}{\sqrt{1-x}} \left(\int_0^{\frac{1}{4}} \frac{1}{1-x\,y}\, \mathrm{d}y\right) \mathrm{d}x. \tag{11.7b}$$

So that, we can use Fubini's theorem and rewrite (11.7b) as:

$$\sum_{n=1}^{\infty} \frac{1}{n^2 \binom{2n}{n}} = \int_0^{\frac{1}{4}} \left(\int_0^1 \frac{1}{(1-x\,y)\sqrt{1-x}}\, \mathrm{d}x\right) \mathrm{d}y. \tag{11.7c}$$

In the inner integral in (11.7c) we change variable $1 - x = \xi^2$ obtaining:

$$\sum_{n=1}^{\infty} \frac{1}{n^2 \binom{2n}{n}} = \int_0^{\frac{1}{4}} \left(\int_0^1 \frac{2}{1-y+y\,\xi^2}\, \mathrm{d}\xi\right) \mathrm{d}y. \tag{11.7d}$$

Finally, what we have obtained in (11.7d) is an elementary integral so that we can proceed with the calculations:

$$\sum_{n=1}^{\infty} \frac{1}{n^2 \binom{2n}{n}} = \int_0^{\frac{1}{4}} \left[\frac{2}{\sqrt{y-y^2}} \arctan\left(\frac{\xi\sqrt{y}}{\sqrt{1-y}}\right) \right]_{\xi=0}^{\xi=1} dy$$

$$= \int_0^{\frac{1}{4}} \frac{2}{\sqrt{y-y^2}} \arctan\left(\frac{\sqrt{y}}{\sqrt{1-y}}\right) \mathrm{d}y. \qquad (11.7e)$$

In (11.7d), we perform the (last) change of variable $\sqrt{y}/\sqrt{1-y} = \eta$, which leads to:

$$\sum_{n=1}^{\infty} \frac{1}{n^2 \binom{2n}{n}} = 4 \int_0^{\frac{1}{\sqrt{3}}} \frac{\arctan \eta}{1+\eta^2} \, \mathrm{d}\eta = 2 \arctan^2\left(\frac{1}{\sqrt{3}}\right),$$

which indeed proves (11.7). □

With a similar technique but much heavier computations, we arrive at the so-called Gosper formula, proved in [1].

Summation 5.

$$\pi = \sum_{n=0}^{\infty} \frac{50n-6}{2^n \binom{3n}{n}}. \qquad (11.8)$$

Proof. Prerequisites chapter 2. We start with the binomial coefficient that appears in (11.8): we have

$$\binom{3n}{n} = \frac{(3n)!}{n!(2n)!} \implies \frac{1}{\binom{3n}{n}} = \frac{(2n)!n!}{(3n)!}.$$

On the other side

$$\mathrm{B}(2n+1, n+1) = \int_0^1 x^{2n}(1-x)^n \mathrm{d}x = \frac{\Gamma(2n+1)\Gamma(n+1)}{\Gamma(3n+2)} = \frac{(2n)!n!}{(3n+1)!}.$$

Therefore

$$\frac{1}{\binom{3n}{n}} = (3n+1) \int_0^1 x^{2n}(1-x)^n \mathrm{d}x. \qquad (11.9)$$

Insert (11.9) on the right-hand side of (11.8); since the exchange between series and integral is again permissible, we arrive at

$$\sum_{n=0}^{\infty}\frac{50n-6}{2^n\binom{3n}{n}}=\sum_{n=0}^{\infty}\frac{(50n-6)(3n+1)}{2^n}\int_0^1 x^{2n}(1-x)^n\mathrm{d}x$$
$$=\int_0^{\infty}\sum_{n=0}^{\infty}(50n-6)(3n+1)\left(\frac{x^2(1-x)}{2}\right)^n\mathrm{d}x.$$

After summing the inner series, which requires a considerable effort in algebraic terms, we see that, again, the sum of the given series is transformed into the calculation of an integral:

$$\sum_{n=0}^{\infty}\frac{50n-6}{2^n\binom{3n}{n}}=\int_0^1\frac{8\left(28x^6-56x^5+28x^4-97x^3+97x^2-6\right)}{\left(x^3-x^2+2\right)^3}\mathrm{d}x.$$

The integral obtained is calculated in terms of elementary functions, but the computational effort is enormous: it must be said that this is an example that the authors of the cited article [1] have identified this integral through a powerful algorithm to produce series having sum π and as a matter of fact, as it can be verified through a software of calculation symbolic it is seen that

$$\int_0^1\frac{8\left(28x^6-56x^5+28x^4-97x^3+97x^2-6\right)}{\left(x^3-x^2+2\right)^3}\mathrm{d}x$$
$$=\left[4\left(\frac{x\left(x^4-29x^3+37x^2-x-8\right)}{\left(x^3-x^2+2\right)^2}-\arctan(1-x)\right)\right]_0^1=\pi.$$

□

Summation 6. Show that

$$\sum_{n=1}^{\infty}(-1)^{n-1}\ln\left(1+\frac{1}{n}\right)=\ln\left(\frac{\pi}{2}\right). \tag{11.10}$$

Proof. Prerequisites chapter 2 and 3. We start observing that

$$\ln\left(1+\frac{1}{n}\right)=\int_0^1\frac{\mathrm{d}x}{x+n}$$

Then, we evaluate (using digamma):

$$s(x)=\sum_{n=1}^{\infty}\frac{(-1)^{n-1}}{n+x}.$$

We have:

$$\begin{aligned} s(x) &= \sum_{n=0}^{\infty} \frac{(-1)^n}{n+1+x} = \sum_{n=0}^{\infty} \left(\frac{1}{2n+x+1} - \frac{1}{2n+x+2} \right) \\ &= \frac{1}{2} \left(\sum_{n=0}^{\infty} \left(\frac{1}{n+\frac{x+1}{2}} - \frac{1}{n+\frac{x+2}{2}} \right) \right) \\ &= \frac{1}{2} \left(\psi\left(\frac{x+2}{2}\right) - \psi\left(\frac{x+1}{2}\right) \right). \end{aligned}$$

Now, returning to the assigned problem:

$$\begin{aligned} \sum_{n=1}^{\infty} (-1)^{n-1} \ln\left(1+\frac{1}{n}\right) &= \sum_{n=1}^{\infty} (-1)^{n-1} \int_0^1 \frac{\mathrm{d}x}{x+n} \\ &= \int_0^1 \sum_{n=1}^{\infty} \frac{(-1)^{n-1}}{x+n} \,\mathrm{d}x = \int_0^1 s(x)\,\mathrm{d}x \\ &= \frac{1}{2} \int_0^1 \left(\psi\left(\frac{x+2}{2}\right) - \psi\left(\frac{x+1}{2}\right) \mathrm{d}x \right) \\ &= \left[\ln\Gamma\left(\frac{x+2}{2}\right) - \ln\Gamma\left(\frac{x+1}{2}\right) \right]_{x=0}^{x=1} \\ &= \ln\left(\Gamma\left(\frac{3}{2}\right) \Gamma\left(\frac{1}{2}\right) \right) \\ &= \ln\left(\frac{\Gamma^2\left(\frac{1}{2}\right)}{2} \right) \\ &= \ln\left(\frac{\pi}{2}\right). \end{aligned}$$

□

The following summation 7 follows from Gauss Theorem 9.4, formula (9.14) and the property (2.15) related to Pochhammer symbols seen at page 63.

Summation 7. Assuming that $m \in \mathbb{N}$, $m > 1$ show that

$$\sum_{n=0}^{\infty} \frac{1}{\binom{n+m}{m}} = \frac{m}{m-1}. \tag{11.11}$$

Proof. Since

$$\binom{n+m}{m} = \frac{(n+m)!}{n!\,m!}.$$

Recalling (2.15) we see that

$$\binom{n+m}{m} = \frac{(m+1)_n}{n!}.$$

Going back on the left-hand side of (11.11), we obtain

$$\begin{aligned}\sum_{n=0}^{\infty}\frac{1}{\binom{n+m}{m}} &= \sum_{n=0}^{\infty}\frac{n!}{(m+1)_n}\\ &= \sum_{n=0}^{\infty}\frac{n!\,n!}{(m+1)_n\,n!}\\ &= \sum_{n=0}^{\infty}\frac{(1)_n\,(1)_n}{(m+1)_n\,n!} &\text{(a)}\\ &= {}_2\mathrm{F}_1\left(\begin{matrix}1;1\\ m+1\end{matrix}\,\middle|\,1\right)\\ &= \frac{\Gamma(m+1)\,\Gamma(m-1)}{\Gamma(m)\,\Gamma(m)}, &\text{(b)}\end{aligned}$$

where (a) follows from the first identity in (2.14), while (b) from Theorem 9.4. The formula (11.11) follows from the functional property of Gamma. Observe that this argument can be extended in the case of $m \in \mathbb{C}$, $\operatorname{Re} m > 1$. □

The next sum requires more advanced methods than those we have seen so far, essentially based on the use of beta. In this case, we will exploit various properties of the hypergeometric function, namely Theorems 9.6, equation (9.27) and 9.5, equation (9.16) discussed in chapter 9. We also need consistent preparatory work to transform semi-factorials into the Pochhammer symbol.

Summation 8.

$$\sum_{n=1}^{\infty}\frac{(4n-1)!!(2n-1)!!}{(4n-2)!!(2n-2)!!}\,\frac{1}{n\,2^n} = \frac{8}{\pi}\sqrt[4]{2}\,\cos\left(\tfrac{\pi}{8}\right)\frac{\Gamma^2\left(\frac{5}{8}\right)}{\Gamma\left(\frac{1}{4}\right)}. \tag{11.12}$$

Proof. Prerequisites chapters 2 and 9. The series on the right-hand side of (11.12) is indeed hypergeometric, once observed that

$$\begin{aligned}&(4n-1)!! = 4^{2n}\left(\tfrac{1}{4}\right)_n\left(\tfrac{3}{4}\right)_n, && (4n-2)!! = 4^{2n-1}\,(n-1)!\,\left(\tfrac{1}{2}\right)_n\\ &(2n-1)!! = 2^2\left(\tfrac{1}{2}\right)_n, && (2n)!! = 2^n\,n!.\end{aligned}$$

We can express the infinite series on the right-hand side of (11.12) as

$$\sum_{n=1}^{\infty}\frac{(4n-1)!!(2n-1)!!}{(4n-2)!!(2n-2)!!}\,\frac{1}{n\,2^n} = 8\sum_{n=1}^{\infty}\frac{\left(\frac{1}{4}\right)_n\left(\frac{3}{4}\right)_n}{n!(n-1)!}\,\frac{1}{2^n}.$$

Let's change the sum index by putting $m = n-1$ since we need a series indexed starting from 0:

$$\sum_{n=1}^{\infty}\frac{(4n-1)!!(2n-1)!!}{(4n-2)!!(2n-2)!!}\,\frac{1}{n\,2^n} = 8\sum_{m=0}^{\infty}\frac{\left(\frac{1}{4}\right)_{m+1}\left(\frac{3}{4}\right)_{m+1}}{(m+1)!m!}\,\frac{1}{2^{m+1}}.$$

Recalling formula (2.16) at page 63 we have

$$\left(\tfrac{1}{4}\right)_{m+1} = \tfrac{1}{4}\left(\tfrac{5}{4}\right)_m\,, \quad \left(\tfrac{3}{4}\right)_{m+1} = \tfrac{3}{4}\left(\tfrac{5}{4}\right)_m\,;$$

moreover since $(2)_m = (m+1)!$ we obtain:

$$\begin{aligned}\sum_{n=1}^{\infty} \frac{(4n-1)!!(2n-1)!!}{(4n-2)!!(2n-2)!!}\,\frac{1}{n\,2^n} &= \frac{3}{4}\sum_{m=0}^{\infty} \frac{\left(\frac{5}{4}\right)_m \left(\frac{7}{4}\right)_m}{(2)_m\, m!}\,\frac{1}{2^m} \\ &= \frac{3}{4}\,{}_2\mathrm{F}_1\left(\begin{matrix}\frac{5}{4};\frac{7}{4}\\ 2\end{matrix}\,\middle|\,\frac{1}{2}\right).\end{aligned}$$

We now invoke the Pfaff identity (9.27) and arrive at:

$$\sum_{n=1}^{\infty} \frac{(4n-1)!!(2n-1)!!}{(4n-2)!!(2n-2)!!}\,\frac{1}{n\,2^n} = \frac{2}{\sqrt[4]{8}}\,{}_2\mathrm{F}_1\left(\begin{matrix}\frac{5}{4};\frac{1}{4}\\ 2\end{matrix}\,\middle|\,-1\right).$$

For the final step we invoke, Kummer Theorem 9.5, which is indeed correct since $1+\frac{5}{4}-\frac{1}{4}=2$. Hence

$$\sum_{n=1}^{\infty} \frac{(4n-1)!!(2n-1)!!}{(4n-2)!!(2n-2)!!}\,\frac{1}{n\,2^n} = \frac{3}{\sqrt[4]{8}}\,\frac{\Gamma\left(\frac{13}{8}\right)}{\Gamma\left(\frac{11}{8}\right)\Gamma\left(\frac{9}{4}\right)}.$$

Or, using the functional property of gamma, equation (2.4) page 59, which is useful since all the gamma arguments we obtained are greater than 1:

$$\sum_{n=1}^{\infty} \frac{(4n-1)!!(2n-1)!!}{(4n-2)!!(2n-2)!!}\,\frac{1}{n\,2^n} = 8\sqrt[4]{2}\,\frac{\Gamma\left(\frac{5}{8}\right)}{\Gamma\left(\frac{1}{4}\right)\Gamma\left(\frac{3}{8}\right)}.$$

At this point, observe that the arguments 3/8 and 5/8 are conjugated; we can use the Euler reflection formula (2.32) to arrive at (11.12). □

Summation 9. Evaluate the double integral below, starting from the x variable:

$$\int_0^{\infty}\int_0^{\infty} \frac{\mathrm{d}x\;\mathrm{d}y}{(1+y)(1+x^2\,y)}\,. \tag{11.13}$$

Then, reverting the order of integration proves that

$$\int_0^{\infty} \frac{\ln x}{x^2-1}\,\mathrm{d}x = \frac{\pi^2}{4}\,. \tag{11.14}$$

Finally, use (11.14), after integration by series, to show that

$$\sum_{n=0}^{\infty} \frac{1}{(2n+1)^2} = \frac{\pi^2}{8}. \tag{11.15}$$

Proof. The argument is taken from [99]. Integrate (11.13) with respect to x, first, and then with respect to y. We obtain, using elementary methods:

$$\int_0^{\infty} \left(\frac{1}{1+y} \int_0^{\infty} \frac{\mathrm{d}x}{1+x^2\,y} \right) \mathrm{d}y = \frac{\pi}{2} \int_0^{\infty} \frac{\mathrm{d}y}{\sqrt{y}(1+y)} = \frac{\pi^2}{2}\,. \tag{11.16}$$

Reverting the order of integration, using a partial fraction decomposition, we obtain

$$\begin{aligned} &\int_0^{\infty} \left(\int_0^{\infty} \frac{\mathrm{d}y}{(1+y)(1+x^2\,y)} \right) \mathrm{d}x \\ &= \int_0^{\infty} \frac{1}{1-x^2} \left(\int_0^{\infty} \left(\frac{1}{1+y} - \frac{x^2}{1+x^2\,y} \right) \mathrm{d}y \right) \mathrm{d}x \\ &= \int_0^{\infty} \frac{1}{1-x^2} \ln \frac{1}{x^2} \mathrm{d}x \;=\; 2 \int_0^{\infty} \frac{\ln x}{x^2-1} \mathrm{d}x\,. \end{aligned} \tag{11.17}$$

Equating (11.16) and (11.17), we get (11.14). For the last step, first, split the integration domain in $[0\,,1]$ plus $[1,\infty)$, and use the change of variable $x = 1/u$ in the integral evaluated on $[1\,,\infty)$:

$$\begin{aligned} \int_0^{\infty} \frac{\ln x}{x^2-1} \mathrm{d}x &= \int_0^{1} \frac{\ln x}{x^2-1} \mathrm{d}x + \int_1^{\infty} \frac{\ln x}{x^2-1} \mathrm{d}x \\ &= \int_0^{1} \frac{\ln x}{x^2-1} \mathrm{d}x + \int_0^{1} \frac{\ln u}{u^2-1} \mathrm{d}u. \end{aligned}$$

So we obtain

$$\int_0^{1} \frac{\ln x}{x^2-1} \mathrm{d}x = \frac{\pi^2}{8}\,. \tag{11.18}$$

Expanding in geometric series:

$$\frac{1}{1-x^2} = \sum_{n=0}^{\infty} x^{2\,n}\,.$$

Then, invoking Theorem 1.8:

$$\int_0^{1} \frac{-\ln x}{1-x^2}\, \mathrm{d}x = \sum_{n=0}^{\infty} \int_0^{1} -x^{2\,n}\, \ln x\, \mathrm{d}x\,.$$

Integrating by parts:

$$\int_0^{1} -x^{2\,n}\, \ln x\, \mathrm{d}x = \left[-\frac{x^{2\,n+1}}{2\,n+1}\, \ln x \right]_0^1 + \int_0^{1} \frac{x^{2\,n}}{2\,n+1}\, \mathrm{d}x = \frac{1}{(2\,n+1)^2}\,.$$

Hence the integral on the left-hand side of (11.18) provides our statement (11.15). □

Remark. From (11.15) we can re-evaluate

$$\zeta(2) = \frac{\pi^2}{6}$$

as presented in (1.30), by means of the identity (4.2) proved at page 125.

Summation 10. Adapting the argument shown in the previous summation 9, evaluate the triple integral

$$I_3 = \iiint_{[0,\,\infty)^3} \frac{\mathrm{d}x\,\mathrm{d}y_1\,\mathrm{d}y_2}{(1+y_1)\,(1+y_2)\,(1+y_1\,y_2\,x^2)}. \tag{11.19}$$

Them use (11.19) to show Euler summation formula:

$$\beta(3) = \sum_{n=0}^{\infty} \frac{(-1)^n}{(2n+1)^3} = \frac{\pi^3}{32}. \tag{11.20}$$

Proof. By integrating (11.19) first in x and then in the variables y_i, we can calculate this integral by elementary methods, the first (arctan) integration being immediate:

$$I_3 = \frac{\pi}{2} \iint_{[0,\,\infty)^2} \frac{\mathrm{d}y_1\,\mathrm{d}y_2}{\sqrt{y_1\,y_2}\,(1+y_1+y_2+y_1\,y_2)}.$$

Then, changing variables in a straightforward way $y_1 = u_1^2$, $y_2 = u_2^2$ we arrive at a second and third elementary integration

$$I_3 = 2\,\pi \iint_{[0,\,\infty)^2} \frac{\mathrm{d}u_1\,\mathrm{d}u_2}{1+u_1^2+u_2^2+u_1^2\,u_2^2} = \pi^2 \int_0^\infty \frac{\mathrm{d}u_2}{1+u_2^2} = \frac{\pi^3}{2}.$$

If we now return to (11.19) and begin by integrating with respect to y_1, after decomposing into partial fractions, we obtain:

$$I_3 = \iint_{[0,\,\infty)^2} \left(\frac{2x^2 \ln x}{(1+x^2)\,(x^2 y_2 - 1)} - \frac{2\ln x}{(1+x^2)\,(1+y_2)} + \frac{\ln y_2}{(1+y_2)\,(x^2 y_2 - 1)} \right) \mathrm{d}x\,\mathrm{d}y_2.$$

It is convenient to organise the double integration differently, in the first two integrals beginning with y_2 in the third, instead of starting with x; in this way, we obtain

$$I_3 = \iint_{[0,\infty)^2} \left(\frac{2x^2 \ln x}{(1+x^2)(x^2 y_2 - 1)} - \frac{2 \ln x}{(1+x^2)(1+y_2)} \right) \mathrm{d}y_2\, \mathrm{d}x$$
$$+ \iint_{[0,\infty)^2} \frac{\ln y_2}{(1+y_2)(x^2 y_2 - 1)} \mathrm{d}x\, \mathrm{d}y_2$$
$$= \int_0^\infty \left[\frac{2 \ln x \ln\left(\frac{x^2 y_2 - 1}{y_2 + 1}\right)}{1+x^2} \right]_{y_2=0}^{y_2 \to \infty} \mathrm{d}x + \int_0^\infty \left[-\frac{\ln y_2 \tanh^{-1}(x\sqrt{y_2})}{\sqrt{y_2}\,(1+y_2)} \right]_{x=0}^{x\to\infty} \mathrm{d}y_2$$
$$= 4 \int_0^\infty \frac{\ln^2 x}{1+x^2} \mathrm{d}x + \int_0^\infty \frac{\ln y_2}{\sqrt{y_2}\,(1+y_2)} \mathrm{d}y_2. \tag{11.21}$$

Now notice that the y_2 integral at right hand side of (11.21) vanishes, indeed:

$$\int_0^\infty \frac{\ln y_2}{\sqrt{y_2}\,(1+y_2)} \mathrm{d}y_2 = \int_0^1 \frac{\ln y_2}{\sqrt{y_2}\,(1+y_2)} \mathrm{d}y_2 + \int_1^\infty \frac{\ln y_2}{\sqrt{y_2}\,(1+y_2)} \mathrm{d}y_2$$
$$= \int_0^1 \frac{\ln y_2}{\sqrt{y_2}\,(1+y_2)} \mathrm{d}y_2 + \int_1^\infty \frac{\ln y_2}{\sqrt{y_2}\,(1+y_2)} \mathrm{d}y_2$$
$$= \int_0^1 \frac{\ln y_2}{\sqrt{y_2}\,(1+y_2)} \mathrm{d}y_2 - \int_0^1 \frac{\ln u}{\sqrt{u}\,(1+u)} \mathrm{d}u = 0.$$

Where we have changed the variable to $y_2 = 1/u$ in the second integral. In conclusion, we have shown that

$$\int_0^\infty \frac{\ln^2 x}{1+x^2} \mathrm{d}x = \frac{\pi^3}{8}. \tag{11.22}$$

Going on, we again split the interval of integration on the left-hand side of (11.22) and change variable $x = 1/u$ arriving at

$$\int_0^\infty \frac{\ln^2 x}{1+x^2} \mathrm{d}x = \int_0^1 \frac{\ln^2 x}{1+x^2} \mathrm{d}x + \int_1^\infty \frac{\ln^2 x}{1+x^2} \mathrm{d}x$$
$$= \int_0^1 \frac{\ln^2 x}{1+x^2} \mathrm{d}x + \int_0^1 \frac{\ln^2 u}{1+u^2} \mathrm{d}u.$$

Finally recalling (11.22), we obtain

$$\int_0^1 \frac{\ln^2 x}{1+x^2} \mathrm{d}x = \frac{\pi^3}{16}. \tag{11.23}$$

Now, expanding in geometric series the denominator of the integrand in (11.23), we find

$$\frac{\pi^3}{16} = \sum_{n=0}^\infty \int_0^1 (-1)^n x^{2n} \ln^2 x \, \mathrm{d}x = 2 \sum_{n=0}^\infty (-1)^n \frac{1}{(2n+1)^3}.$$

Hence, we proved (11.20). □

11.2 Definite integrals

One of the most fruitful fields for special functions is the calculation of definite integrals. Beyond the theoretical aspect of establishing identities, this allows, on the one hand, the validation of the numerical approximation techniques of the considered integral and, on the other, the testing of the accuracy of evaluating the special functions. Furthermore, each of the examples we will bring to the reader's attention allows us to consolidate and appreciate the theoretical structure of what we presented in the previous chapters.

Definite integral 1.

$$\int_0^1 \frac{1}{\sqrt{x}} \arctan\left(\frac{1+\sqrt{1-x}}{1+\sqrt{1+x}}\right) dx = \frac{\pi}{4} + \frac{1}{\sqrt{2\,\pi}}\,\Gamma^2\left(\frac{3}{4}\right) \tag{11.24}$$

$$\int_0^1 \sqrt{x} \arctan\left(\frac{1+\sqrt{1-x}}{1+\sqrt{1+x}}\right) dx = \frac{\pi}{12} + \frac{1}{36\sqrt{2\,\pi}}\,\Gamma^2\left(\frac{1}{4}\right) \tag{11.25}$$

Proof. Prerequisites Theorem 1.19 of chapters 1 and 2, Theorem 2.9. Both integrals (11.24) and (11.25) are evaluated using the parametric integral

$$J(a) = \int_0^1 x^a \arctan\left(\frac{1+\sqrt{1-x}}{1+\sqrt{1+x}}\right) dx.$$

After an integration by parts, we see that

$$\begin{aligned} J(\alpha) &= \frac{1}{(1+\alpha)} \arctan(\sqrt{2}-1) + \frac{1}{4(1+\alpha)} \int_0^1 \frac{x^{1+\alpha}}{\sqrt{1-x^2}}\, dx \\ &= \frac{\pi}{8(1+\alpha)} + \frac{1}{8(1+\alpha)} \int_0^1 x^{\frac{\alpha}{2}}(1-x)^{-\frac{1}{2}}\, dx \\ &= \frac{\pi}{8(1+\alpha)} + \frac{1}{8(1+\alpha)}\, \mathrm{B}\left(\frac{\alpha}{2}+1, \frac{1}{2}\right). \end{aligned}$$

Formulæ (11.24) and (11.25) follow taking $\alpha = \pm\frac{1}{2}$ and recalling Theorem 2.9. □

Definite integral 2. If $0 < x < 2$

$$\int_0^{+\infty} \frac{t^x - 1}{(1+t^3)\ln t}\, dt = \ln\left[\tan\left(\frac{\pi}{6}(1+x)\right)\right] + \ln\sqrt{3}. \tag{11.26}$$

Proof. Prerequisites Theorem 1.19 of chapters 1 and 2. We can differentiate the parametric integral, and then after a change of variable, we obtain

$$F'(x) = \int_0^{+\infty} \frac{t^x}{1+t^3}\, dt = \frac{1}{3}\int_0^{+\infty} \frac{s^{\frac{x}{3}-\frac{2}{3}}}{1+s}\, ds.$$

Recalling (2.33) at page 73 we arrive at

$$F'(x) = \frac{\pi}{3}\,\frac{1}{\sin\left[\frac{\pi}{3}\,(x+1)\right]}.$$

Hence integrating

$$F(x) = \frac{\pi}{3}\int_0^x \frac{1}{\sin\left[\frac{\pi}{3}\,(u+1)\right]}\,\mathrm{d}u = \int_{\frac{\pi}{3}}^{\frac{\pi}{3}\,(1+x)} \frac{\mathrm{d}s}{\sin s}.$$

Now, recalling

$$\int \frac{\mathrm{d}s}{\sin s} = \ln\tan\frac{s}{2}.$$

we arrive at (11.26).

□

Definite integral 3. Show that

$$\int_0^1 \frac{\mathrm{d}x}{\sqrt{x^4-3x^2+2}} = \frac{1}{4\sqrt{2}\,\pi}\,\Gamma^2\left(\frac{1}{4}\right).$$

Proof. Prerequisites chapters 2 and 9.

$$\int_0^1 \frac{\mathrm{d}x}{\sqrt{x^4-3x^2+2}} = \frac{1}{2}\int_0^1 \frac{\mathrm{d}u}{\sqrt{u(1-u)(2-u)}} \tag{a}$$

$$= \frac{1}{2\sqrt{2}}\int_0^1 \frac{u^{-1/2}(1-u)^{-1/2}}{\sqrt{1-\frac{1}{2}\,u}}\,\mathrm{d}u. \tag{b}$$

Notice that in (a), after the factorisation $x^4-3x^2+2=(1-x^2)(2-x^2)$, we change variable $x^2=u$ while in (b), we set up the integral to apply the integral representation (9.8) arriving to

$$\int_0^1 \frac{\mathrm{d}x}{\sqrt{x^4-3x^2+2}} = \frac{\pi}{2\sqrt{2}}\,{}_2\mathrm{F}_1\left(\left.\begin{matrix}\frac{1}{2},\frac{1}{2}\\ 1\end{matrix}\,\right|\,\frac{1}{2}\right).$$

Now, recalling identity (9.20) page 203 taking $a=1/2$ and $c=1$ we see that:

$$\int_0^1 \frac{\mathrm{d}x}{\sqrt{x^4-3x^2+2}} = \frac{\pi}{2\sqrt{2}}\,\frac{\Gamma(\frac{1}{2})\,\Gamma(1)}{\Gamma^2(\frac{3}{4})}.$$

Using the reflection formula (2.32), the thesis follows. □

Definite integral 4. If $0<\alpha<1$

$$\int_0^1 \frac{x^\alpha - x^{-\alpha}}{1-x}\,\mathrm{d}x = \frac{\pi}{\tan(\alpha\,\pi)} - \frac{1}{\alpha}. \tag{11.27}$$

Proof. Prerequisites chapter 3. Using the geometric series expansion

$$\frac{1}{1-x} = \sum_{n=1}^{\infty} x^{n-1}, \quad |x| < 1,$$

we obtain:

$$\begin{aligned}\int_0^1 \frac{x^\alpha - x^{-\alpha}}{1-x}\, \mathrm{d}x &= \sum_{n=1}^{\infty} \int_0^1 \left(x^{n-1+\alpha} - x^{n-1-\alpha}\right) \\ &= \sum_{n=1}^{\infty} \left(\frac{1}{n+\alpha} - \frac{1}{n-\alpha}\right).\end{aligned}$$

Recalling the representation of digamma given in Theorem 3.1 equation, page 108 equation (3.1) we obtain:

$$\begin{aligned}\int_0^1 \frac{x^\alpha - x^{-\alpha}}{1-x}\, \mathrm{d}x &= \sum_{n=1}^{\infty} \left(\frac{1}{n+\alpha} - \frac{1}{n} + \frac{1}{n} - \frac{1}{n-\alpha}\right) \\ &= \sum_{n=1}^{\infty} \left(\frac{1}{n} - \frac{1}{n-1-(\alpha-1)}\right) - \sum_{n=1}^{\infty} \left(\frac{1}{n} - \frac{1}{n-1+(\alpha+1)}\right).\end{aligned}$$

Hence

$$\int_0^1 \frac{x^\alpha - x^{-\alpha}}{1-x}\, \mathrm{d}x = \psi(\alpha-1) - \psi(\alpha+1).$$

Eventually recalling the functional characterisation of $\psi(x)$ formula (3.3) we arrive at

$$\int_0^1 \frac{x^\alpha - x^{-\alpha}}{1-x}\, \mathrm{d}x = \psi(\alpha-1) - \psi(\alpha) - \frac{1}{\alpha}.$$

The last step to obtain (11.27) comes from the reflection formula (3.7). □

Definite integral 5. Show that

$$\int_0^{+\infty} \frac{\left(\sqrt{1+x^2} - x\right) \ln x}{\sqrt{x}}\, \mathrm{d}x = -\frac{2}{9\sqrt{\pi}} \Gamma^2\left(\frac{1}{4}\right). \tag{11.28}$$

Proof. Consider the parametric integral

$$I(a) = \int_0^{+\infty} x^a \left(\sqrt{1+x^2} - x\right) \mathrm{d}x$$

and, then, observe that integral (11.28) is given by $I'(-1/2)$. Since $\sqrt{1+x^2} - x = \frac{1}{2x} + o(x^{-1})$ for $x \to +\infty$, $I(a)$ is defined for $-1 < a < 0$. The (Euler) substitution $x = (1-t^2)/(2t)$ leads to

$$I(a) = \frac{1}{2^{a+1}} \int_0^1 t^{-(a+1)} \left(1+t^2\right) \left(1-t^2\right)^a \mathrm{d}t.$$

Hence, using the change of variable $t^2 = u$, we obtain, due to Theorem 9.3

$$I(a) = \frac{1}{2^{a+2}} \int_0^1 \frac{(1-u)^a u^{-\frac{a}{2}-1}}{1+u} \, du$$
$$= \frac{1}{2^{a+2}} \frac{a\,\Gamma\left(-\frac{a}{2}\right)\Gamma(a)}{\Gamma\left(\frac{a+2}{2}\right)} \, {}_2\mathrm{F}_1\left(\left.\begin{matrix} -\frac{a}{2}, -1 \\ 1+\frac{a}{2} \end{matrix}\,\right|\, -1\right).$$

Eventually, since one of the ${}_2\mathrm{F}_1$ parameters is a negative integer, we invoke Theorem 9.1, and after some simplification, and also calling the Legendre duplication formula (2.50), we arrive at

$$I(a) = \frac{\Gamma\left(-\frac{a}{2}\right)\Gamma(a)}{2^a(a+2)\Gamma\left(\frac{a}{2}\right)} = \frac{\Gamma\left(-\frac{a}{2}\right)\Gamma\left(\frac{a}{2}+\frac{1}{2}\right)}{2(a+2)\sqrt{\pi}}.$$

For the second part, observe that

$$I'(a) = \frac{\Gamma\left(-\frac{a}{2}\right)\Gamma\left(\frac{a+1}{2}\right)\left((a+2)\psi\left(\frac{a+1}{2}\right) - (a+2)\psi\left(-\frac{a}{2}\right) - 2\right)}{4(a+2)^2\sqrt{\pi}}.$$

Notice that we used the differentiation rule $\Gamma'(z) = \Gamma(z)\,\psi(z)$. Moreover, the term $(a+2)\psi\left(\frac{a+1}{2}\right) - (a+2)\psi\left(-\frac{a}{2}\right)$ vanishes for $a = -\frac{1}{2}$ while

$$a = -\frac{1}{2} \implies \Gamma\left(-\frac{a}{2}\right)\Gamma\left(\frac{a+1}{2}\right) = \Gamma^2\left(\frac{1}{4}\right)$$

and then (11.28) is proved. □

Definite integral 6.

$$\int_0^1 \frac{dx}{\sqrt[4]{(1-x)(1+x)^3}} = \frac{1}{\sqrt{2}}\left(\pi + \ln\left(3-2\sqrt{2}\right)\right). \tag{11.29}$$

Proof. Prerequisites chapters 2, 3, and 9. On the left-hand side, the integral change variable $1 - x = u$ leads to

$$\int_0^1 \frac{dx}{\sqrt[4]{(1-x)(1+x)^3}} = \int_0^1 \frac{u^{-\frac{1}{4}}}{(2-u)^{3/4}} \, du = \frac{1}{2^{\frac{3}{4}}} \int_0^1 \frac{u^{-\frac{1}{4}}}{(1-\frac{1}{2}u)^{3/4}} \, du.$$

We can now apply Theorem 9.3 and recognise the hypergeometric nature of the given integral; in fact

$$\int_0^1 \frac{dx}{\sqrt[4]{(1-x)(1+x)^3}} = \frac{1}{2^{\frac{3}{4}}} \frac{\Gamma\left(\frac{3}{4}\right)}{\Gamma\left(\frac{7}{4}\right)} \, {}_2\mathrm{F}_1\left(\left.\begin{matrix} \frac{3}{4}; \frac{3}{4} \\ \frac{7}{4} \end{matrix}\,\right|\, \frac{1}{2}\right)$$

Now we use the reflection formula, see page 72, and Euler transformation, Theorem 9.7, equation (9.28) to arrive at

$$\int_0^1 \frac{dx}{\sqrt[4]{(1-x)(1+x)^3}} = \frac{2}{3} \, {}_2\mathrm{F}_1\left(\left.\begin{matrix} 1; 1 \\ \frac{7}{4} \end{matrix}\,\right|\, \frac{1}{2}\right).$$

Hence, we can apply the formula (9.21) and infer that

$$\int_0^1 \frac{\mathrm{d}x}{\sqrt[4]{(1-x)(1+x)^3}} = \frac{1}{2}\left(\psi\left(\frac{7}{8}\right) - \psi\left(\frac{3}{8}\right)\right).$$

At this point, Gauss formula (3.39), Theorem 3.11 for the rational values of digamma, which is implemented in major computer algebra systems, like Mathematica®, leads to (11.29). □

Definite integral 7. Show that, if $a > 1$:

$$\int_1^\infty \frac{\arctan x}{x^a}\,\mathrm{d}x = \frac{1}{4(a-1)}\left(\pi + \psi\left(\frac{a+2}{4}\right) - \psi\left(\frac{a}{4}\right)\right). \tag{11.30}$$

This problem appeared in the Romanian Mathematical Magazine.

Proof. Prerequisites chapter 2, and chapter 3, Theorem 3.1. First, we normalise the interval of integration:

$$\begin{aligned}
\int_1^\infty \frac{\arctan x}{x^a}\,\mathrm{d}x &= \int_0^1 x^{\alpha-2}\arctan\frac{1}{x}\,\mathrm{d}x \\
&= \int_0^1 x^{\alpha-2}\left(\frac{\pi}{2} - \arctan x\right)\mathrm{d}x && \text{(a)} \\
&= \frac{\pi}{2(\alpha-1)} - \int_0^1 x^{\alpha-2}\arctan x\,\mathrm{d}x && \text{(b)} \\
&= \frac{\pi}{2(\alpha-1)} - \left(\frac{\pi}{4(\alpha-1)} - \frac{1}{\alpha-1}\int_0^1 \frac{x^{\alpha-1}}{1+x^2}\,\mathrm{d}x\right) && \text{(c)} \\
&= \frac{\pi}{4(\alpha-1)} + \frac{1}{2(\alpha-1)}\int_0^1 \frac{x^{\frac{\alpha}{2}-1}}{1+x}\,\mathrm{d}x. && \text{(d)}
\end{aligned}$$

In (a), we used a property of arctan function; in (b), we evaluate the monomial integral; in (c), we integrate by parts; and in (d), we changed the variable of integration. Now we focus on the integral on the right-hand side of (d), representing the decominator as the sum of a geometric series:

$$\int_0^1 \frac{x^{\frac{\alpha}{2}-1}}{1+x}\,\mathrm{d}x = \sum_{n=0}^\infty (-1)^n \int_0^1 x^{n+\frac{\alpha}{2}-1}\mathrm{d}x = \sum_{n=0}^\infty \frac{(-1)^n}{n+\frac{\alpha}{2}}.$$

Hence, separating odd end even indexes, we get

$$\begin{aligned}
\int_0^1 \frac{x^{\frac{\alpha}{2}-1}}{1+x}\,\mathrm{d}x &= \sum_{n=0}^\infty \left(\frac{1}{2n+\frac{\alpha}{2}} - \frac{1}{2n+1+\frac{\alpha}{2}}\right) \\
&= \frac{1}{2}\sum_{n=0}^\infty \left(\frac{1}{n+\frac{\alpha}{4}} - \frac{1}{n+\frac{1}{2}+\frac{\alpha}{4}}\right) \\
&= \frac{1}{2}\left(\psi\left(\frac{\alpha+2}{4}\right) - \psi\left(\frac{\alpha}{4}\right)\right). && \text{(e)}
\end{aligned}$$

Notice that (e) stems from Theorem 3.1, ending our proof. □

Bibliography

[1] G. Almkvist, C. Krattenthaler, and J. Petersson. Some new formulas for π. *Experimental Mathematics*, 12(4):441–456, 2003.

[2] H. Alzer. On an inequality of H. Minc and L. Sathre. *Journal of Mathematical Analysis and Applications*, 179:396–402, 1993.

[3] H. Alzer. On some inequalities for the gamma and psi functions. *Mathematics of Computation*, 66(217):373–389, 1997.

[4] G.E. Andrews, R. Askey, and R. Roy. *Special Functions.* Cambridge University press, Cambridge, 1999.

[5] G.E. Andrews, R. Askey, and R. Roy. *Special Functions.* Cambridge University Press, Cambridge, 1999.

[6] L.C. Andrews. *Special Functions of Mathematics for engineers. Second Edition.* Oxford University press, Oxford, 1998.

[7] C. Apelian and S. Surace. *Real and Complex Analysis.* CRC Press, Boca Raton, 2010.

[8] R. Apéry. Irrationalité de $\zeta(2)$ et $\zeta(3)$. *Astérisque*, 61(11-13):1, 1979.

[9] T.M. Apostol. *Analytic number theory.* Springer, New York, 1976.

[10] Tom M. Apostol. *Calculus: Multi Variable Calculus and Linear Algebra, with applications to Differential Equations and Probability.* John Wiley & Sons, New York, U.S.A., 1969.

[11] P. Appell and J. Kampè de Fèriet. *Fonctions Hypergéometriques et Hypersphériques; Polynomes d'Hermite.* Gautier-Villars, Paris, 1926.

[12] G.B. Arfken, H.J. Weber, and F.E. Harris. *Mathematical Methods for Physicists (Seventh Edition).* Academic Press, Boston, 2013.

[13] A.K. Arora, S.K. Goel, and D.M. Rodriguez. Special integration techniques for trigonometric integrals. *The American Mathematical Monthly*, 95(2):126–130, 1988.

[14] Emil Artin and Michael Butler. *The Gamma function.* Holt, Rinehart and Winston, New York, U.S.A., 1964.

[15] C. Arzelà. Sulla integrazione per serie. *Atti Acc. Lincei Rend.*, 5(6):290–292, 1885.

[16] A. Aycock. Euler and the multiplication formula for the Gamma Function. *Euleriana*, 1(2):197–204, 2021.

[17] R. Ayoub. Euler and the Zeta function. *The American Mathematical Monthly*, 81(10):1067–1086, 1974.

[18] W.N. Bailey. *Generalized hypergeometric series.* Cambridge University Press, Cambridge, 1935.

[19] H. Bateman and A. Erdélyi. *Higher transcendental functions, Volume 1.* McGraw-Hill, New York, 1955.

[20] D. Bernoulli. *Exercitationes quaedam mathematicæ.* Apud Dominicum Lovisam, Venice, 1724.

[21] R. Bhatia. *Fourier series*, volume 5. MAA, Providence, 2005.

[22] H.A. Bohr and J. Mollerup. *Laerebog i matematisk analyse af Harald Bohr og Johannes Mollerup*, volume 3. J. Gjellerups, 1922.

[23] P. Borwein, S. Choi, B. Rooney, and A. Weirathmueller. *The Riemann hypothesis: a resource for the Afficionado and Virtuoso alike.* Springer, New York, 2008.

[24] F. Bowman. *Introduction to Bessel Functions.* Dover, New York, 1958.

[25] F. Bowman. *Introduction to Elliptic Functions.* Dover, New York, 1961.

[26] J.G. Brida, G. Mingari Scarpello, and D. Ritelli. The Solow model with logistic manpower: a stability analysis. *The Journal of World Economic Review*, 49:161–166, 2008.

[27] J.W. Brown and R.V. Churchill. *Fourier Series and Boundary Value Problems. Seventh ed.* McGraw–Hill, New York, 2008.

[28] Paul F. Byrd and Morris D. Friedman. *Handbook of elliptic integrals for engineers and scientists.* Springer, New York, 1971.

[29] B.C. Carlson. *Special functions of applied mathematics.* Academic Press, New York, 1977.

[30] A.L. Cauchy. Mémoire sur les intégrales définies, prises entre des limites imaginaires. *Bulletin des sciences mathématiques et astronomiques*, 7(1):265–304, 1874.

[31] A. Chakrabarti. *Elements of Ordinary Differential Equations and Special Functions.* John Wiley & Sons, New York, 2006.

[32] H. Chen. Evaluation of the Laplace integral. *International Journal of Mathematical Education in Science and Technology*, 35(5):773–777, 2004.

[33] F. Chorlton. Studies in Bessel functions via Laplace transforms. *International Journal of Mathematical Education in Science and Technology*, 29(3):437–473, 1998.

[34] E.A. Coddington and N. Levinson. *Theory of ordinary differential equations.* McGraw-Hill, New York, 1955.

[35] D.F. Connon. A new derivation of Hermite's integral for the Hurwitz zeta function. *arXiv preprint arXiv:0908.1427*, 2009.

[36] V.A. Dobrushkin. *Applied Differential Equations with Boundary Value Problems.* CRC press, Boca Raton, 2018.

[37] Vladimir A. Dobrushkin. *Applied differential equations: The primary course.* CRC Press, Boca Raton, 2022.

[38] G. Doetsch. *Introduction to the Theory and Application of the Laplace Transformation.* Springer, New York, 1974.

[39] P.L. Duren. *Invitation to Classical Analysis*, volume 17. American Mathematical Society, Providence, 2012.

[40] J. Dutka. The incomplete beta function - a historical profile. *Archive for History of Exact Sciences*, 324(1):11–29, 1981.

[41] J. Dutka. The Early History of the Hypergeometric Function. *Archive for History of Exact Sciences*, 31(1):15–34, 1984.

[42] J. Dutka. The Early History of the Factorial Function. *Archive for History of Exact Sciences*, 34(3):225–249, 1991.

[43] J. Dutka. On the Early History of Bessel Functions. *Archive for History of Exact Sciences*, 49(2):105–134, 1995.

[44] P.P.G. Dyke. *An introduction to Laplace transforms and Fourier series.* Springer, New York, 2001.

[45] H.M. Edwards. *Riemann's zeta function.* Academic press, New York, 1974.

[46] J. Edwards. *A treatise on the integral calculus with application and Examples, vol.2.* Chelsea Pulishing Company, New York, U.S.A., 1922.

[47] V.A. Erdélyi. Der zusammenhang zwischen verschie-denen integraldarstellungen hypergeometrischer funktionen. *The Quarterly Journal of Mathematics*, 8:200–213, 1937.

[48] L. Euler. De summis serierum reciprocarum. *Commentarii Academiæ Scientiarum Petropolitanae*, 7:123–134, 1740.

[49] L. Euler. De summis serierum numeros Bernoullianos involventium. *Novi Commentarii Academiæ Scientiarum Petropolitanae*, 14:129–167, 1769.

[50] O.J. Farrell and B. Ross. *Solved problems: gamma and beta functions, Legendre polynomials, Bessel functions.* Macmillan, New York, 1963.

[51] A. Gambini, G. Mingari Scarpello, and D. Ritelli. Mathematical properties of EOQ models with special cost structure. *Applied Mathematical Modelling*, 37(3):659–666, 2013.

[52] K. F. Gauss. *Gesammelte Werke, vol. 3.* Teubner, Leipzig, 1823.

[53] W. Gautschi. Some elementary inequalities relating to the gamma and incomplete gamma function. *Journal of Mathematical Physics*, 38(1):77–81, 1959.

[54] S. Gelbart and S. Miller. Riemann's zeta function and beyond. *Bulletin of the American Mathematical Society*, 41(1):59–112, 2004.

[55] A. Ghizzetti, A. Ossicini, and L. Marchetti. *Lezioni di complementi di matematica.* 2nd ed. Libreria eredi Virgilo Veschi, Rome, 1972.

[56] L. Glasser, K.T. Kohl, C. Koutschan, V.H. Moll, and A. Straub. The integrals in Gradshteyn and Ryzhik. part 22: Bessel-k functions. *Scientia. Series A. Mathematical Sciences. New Series*, 22:129–151, 2012.

[57] L. Gordon. A stochastic approach to the gamma function. *The American Mathematical Monthly*, 101(9):858–865, 1994.

[58] Izrail S. Gradshteyn and Jossif M. Ryzhik. *Table of Integrals, Series and Products* 6th ed. Academic Press, New York, 2000.

[59] G. Grimmett and D. Stirzaker. *One thousand exercises in probability.* Oxford University Press, Oxford, 2020.

[60] G. Grimmett and D. Stirzaker. *Probability and random processes.* Oxford University Press, Oxford, 2020.

[61] R. Groux and P. Soulat. *Les fonctions spéciales vues par les problèmes.* Cépadues, Toulose, 2010.

[62] P.R. Halmos. *Naive set theory.* van Nostrand, New York, 1960.

[63] O. Hijab. *Introduction to Calculus and Classical Analysis. 2nd Edition.* Springer, New York, 2007.

[64] E. Hille. *Ordinary differential equations in the complex domain.* John Wiley & Sons, New York, 1976.

[65] J.M. Howie. *Complex analysis.* Springer, New York, 2003.

[66] A. Ivić. *The Riemann zeta-function: Theory and applications.* John Wiley & Sons, New York, 1985.

[67] J. Jacod and P. Protter. *Probability essentials.* Springer, New York, 2004.

[68] J. Jensen. An Elementary Exposition of the Theory of the Gamma function. *The Annals of Mathematics*, 17(3):124–166, 1916.

[69] J.P. Juchem Neto, G. Mingari Scarpello, D. Ritelli, and J.C. Claeyssen. Closed–form solution for the Solow model with constant migration. *TEMA*, 16:1–33, 2015.

[70] A.A. Karatasuba and S.M. Voronin. *The Riemann zeta-function.* Walter de Gruyter, Berlin, 1992.

[71] D.E. Knuth. Two notes on notation. *American Mathematical Monthly*, 99(5):403–422, 1992.

[72] T.W. Körner. *Fourier analysis.* Cambridge University Press, Cambridge, U.K., 1989.

[73] P.S. Laplace. *Théorie analytique des probabilités.* Courcier, Paris, 1814.

[74] N.N. Lebedev and R.A. Silvermann. *Special Functions and their applications.* Prentice Hall, New York, 1965.

[75] A.M. Legendre. *Traité des fonctions elliptiques et des intégrales Euleriennes*, volume 2. Huzard-Courcier, Paris, 1826.

[76] J. Liouville. Remarques nouvelles sur l'equation de Riccati. *Journal de mathématiques pures et appliquées*, 6:1–13, 1841.

[77] P. Loya. *Amazing and aesthetic aspects of analysis.* Springer, New York, 2006.

[78] W.A.J. Luxemburg. Arzelà's dominated convergence theorem for the Riemann integral. *The American Mathematical Monthly*, 78(9):970–979, 1971.

[79] C.C. Maican. *Integral evaluations using the Gamma and Beta functions and elliptic integrals in engineering: A Self-study Approach.* International Press of Boston Inc., Boston, 2005.

[80] A.M. Mathai and H.J. Haubold. *Special functions for applied scientists.* Springer, New York, 2008.

[81] H. Minc and L. Sathre. Some inequalities involving $(r!)^{1/r}$. *Proceedings of the Edinburgh Mathematical Society*, 14(1):41–46, 1964.

[82] G. Mingari Scarpello and D. Ritelli. The Solow model improved through the logistic manpower growth law. *Annali Università di Ferrara - Sez VII - Sc. Mat.*, IL:73–83, 2003.

[83] G. Mingari Scarpello and D. Ritelli. On computing some special values of multivariate hypergeometric functions. *Journal of Mathematical Analysis and Applications*, 420:1693–1718, 2014.

[84] F. Monna and G. Monna. *Suites et Séries de Fonctions.* Cépaduès, Toulouse, 2013.

[85] P. J. Nahin. *Inside Interesting Integrals. 2nd Edition.* Springer, New York, 2020.

[86] G. Nemes. On the coefficients of the asymptotic expansion of $n!$. *Journal of Integer Sequences*, 13(6):Article 10.6.6, 5 pp, 2010.

[87] N. Nielsen. *Handbuch der Theorie der Gammafunktion.* Teubner, Leipzg, 1906.

[88] A.S. Olaikhan. *An Introduction to the Harmonic Series and Logarithmic Integrals: For High School Students up to Researchers.* Amazon Digital Services, 2021.

[89] B. Osgood. *Lectures on the Fourier transform and its applications.* American Mathematical Society, Providence, Rhode Island, U.S.A., 2019.

[90] M.S. Paolella. *Intermediate probability: A computational approach.* John Wiley & Sons, New York, 2007.

[91] J.M. Patin. A very short proof of Stirling's formula. *American Mathematical Monthly*, 96(1):41–42, 1989.

[92] B. Patra. *Complex Varables and Special Functions.* PHI Learning, Delhi, 2014.

[93] H.A. Priestley. *Introduction to Complex Analysis 2nd Edition.* Oxford University Press, Oxford, 2003.

[94] M.H. Protter and C.B. Morrey. *Intermediate calculus, 2nd Edition.* Springer, New York, 1985.

[95] E.D. Rainville and P.E. Bedient. *Elementary differential equations. 6th Edition.* Macmillan, New York, 1981.

[96] E.D. Rainville. *Special functions.* Chelsea, New York, 1960.

[97] E.D. Rainville. *Intermediate differential equations.* Macmillan, New York, 1964.

[98] B. Riemann. Ueber die Anzahl der Primzahlen unter einer gegebenen Grösse. *Monatsberichte der Königlich Preuß ischen Akademie der Wissenschaften zu Berlin*, pages 136–144, 1859.

[99] D. Ritelli. Another proof of $\zeta(2) = \frac{\pi^2}{6}$ using double integrals. *American Mathematical Monthly*, 120:642–645, 2013.

[100] D. Ritelli. *Lezioni di Analisi Matematica, quarta edizione.* Esculapio, Bologna, 2021.

[101] D. Ritelli and G. Spaletta. *Introductory Mathematical Analysis for Quantitative Finance.* Chapman and Hall, Boca Raton, 2020.

[102] W. Rogosinski. *Fourier Series.* Chelsea Publishing Company, New York, 1950.

[103] W. Rudin. *Principles of mathematical analysis. 3rd Edition* McGraw–Hill, New York, U.S.A., 1976.

[104] J.B. Seaborn. *Hypergeometric functions and their applications.* Springer, New York, 1991.

[105] L.J. Slater. *Generalized hypergeometric functions.* Cambridge University Press, Cambridge, 1966.

[106] M.R. Spiegel. Some Interesting Special Cases of the Hypergeometric Series. *American Mathematical Monthly*, 69:894–896, 1962.

[107] M.R. Spiegel. *Theory and Problems of Laplace Transforms.* McGraw-Hill, New York, 1965.

[108] M.Z. Spivey. *The art of proving binomial identities.* CRC Press, Boca Raton, 2019.

[109] E.C. Titchmarsh and D.R. Heath-Brown. *The theory of the Riemann zeta-function.* Oxford university press, Oxford, 1986.

[110] O. Vallee and M. Soares. *Airy functions and applications to physics.* World Scientific, Singapore, 2010.

[111] C. Viola. *An Introduction to Special Functions.* Springer, New York, 2016.

[112] W.R. Wade. *An introduction to analysis. 3rd Edition* Pearson Education, New York, U.S.A., 2004.

[113] G.N. Watson. *A treatise on the Theory of the Bessel Functions. Second edition.* Cambridge University press, Cambridge, 1995.

[114] E.T. Whittaker and G.N. Watson. *A Course of Modern Analysis, 5th revised ed. by Victor H. Moll.* Cambridge University press, Cambridge, 2021.

[115] R.C. Wrede and M.R. Spiegel. *Schaum's Outline of advanced calculus.* McGraw–Hill, New York, 2010.

[116] N. Young. *An introduction to Hilbert space.* Cambridge University Press, Cambridge, 1988.

Analytic Index

Kolibri 360 GmbH
Ferdinand-Jühlke-Straße 7
99095 Erfurt, Deutschland
produktsicherheit@kolibri360.de